Big Data Analytics for
Connected Vehicles and Smart Cities

Big Data Analytics for Connected Vehicles and Smart Cities

Bob McQueen

ARTECH
HOUSE

BOSTON | LONDON
artechhouse.com

Library of Congress Cataloging-in-Publication Data
A catalog record for this book is available from the U.S. Library of Congress.

British Library Cataloguing in Publication Data
A catalogue record for this book is available from the British Library.

Cover design by John Gomes

ISBN 13: 978-1-63081-321-5

© 2017 ARTECH HOUSE
685 Canton Street
Norwood, MA 02062

10 9 8 7 6 5 4 3 2 1

Contents

Preface   *xv*

1   **Introduction**   **1**

1.1   Introduction   1

1.2   Informational Objectives of This Chapter   1

1.3   Chapter Word Cloud   2

1.4   Background   2

1.5   Why This Subject and Why Now?   4

1.6   Intended Readership Groups for the Book   5

1.7   Overview of Contents   6

     References   12

2   **Questions to Be Addressed**   **13**

2.1   Informational Objectives of This Chapter   13

2.2   Chapter Word Cloud   13

2.3   Introduction   13

2.4	Questions Instead of Answers	15
2.5	Overview of the Questions	15
2.6	Safety-Related Questions	20
2.7	Efficiency-Related Questions	21
2.8	User Experience-Related Questions	27
2.9	What Do We Do with the Questions?	29
	References	29
3	**What Is Big Data?**	**31**
3.1	Informational Objectives of This Chapter	31
3.2	Chapter Word Cloud	31
3.3	Introduction	31
3.4	How Is Big Data Measured?	32
3.5	What Is Big Data?	33
3.6	Challenges	42
3.7	Big Data in Transportation	46
3.8	Transportation Systems Management and Operations	51
	References	53
4	**Connected and Autonomous Vehicles**	**55**
4.1	Informational Objectives	55
4.2	Chapter Word Cloud	55
4.3	Introduction	56
4.4	What Is a Connected Vehicle?	58
4.5	Connected Vehicle Challenges	60
4.6	What Is an Autonomous Vehicle?	67

4.7	Autonomous Vehicle Challenges	69
4.8	Summary of the Differences between Connected and Autonomous Vehicles	72
4.8	Connected and Autonomous Vehicles within a Smart City	73
4.9	The Likely Impact of the Connected and the Autonomous Vehicle on Transportation	74
4.10	Big Data and Connectivity	75
4.11	Connected and Autonomous Vehicles within a Smart City	75
4.12	The Likely Effect of Connected and Autonomous Vehicles on the Automotive Industry	77
4.13	Summary	79
	References	80
5	**Smart Cities**	**81**
5.1	Informational Objectives	81
5.2	Chapter Word Cloud	81
5.3	Introduction	81
5.4	What Is a Smart City?	83
5.5	Smart City Objectives	91
5.6	Steps Toward a Smart City	92
5.7	Smart City Frameworks	98
5.8	Evaluating the Effects of Investments	104
5.9	Smart City Challenges	104
5.10	Smart City Opportunities	106
5.11	Lessons Learned from the London Congestion Charge Project	108

5.12	The Sentient City	113
5.13	Summary	114
	References	114

6	**What Are Analytics?**	**117**
6.1	Informational Objectives	117
6.2	Introduction	117
6.3	Chapter Word Cloud	118
6.4	What Is an Analytic?	119
6.5	Why Analytics Are Valuable	120
6.6	Smart City Services Analytics	122
6.7	Analytical Performance Management for a Smart City	132
6.8	How Do Analytics and Data Lakes Fit Together?	133
6.9	How to Identify Data Needs Associated with Analytics	134
6.10	Summary	134
	References	135

7	**The Practical Application of Analytics to Transportation**	**137**
7.1	Informational Objectives of This Chapter	137
7.2	Chapter Word Cloud	138
7.3	Introduction	138
7.4	Integrated Payment Systems—What Are They?	139
7.5	Why Does Integrated Payment Make a Good Departure Point for a Smart City?	140
7.6	Integrated Payment System Analytics and Their Practical Application	141
7.7	MaaS—What Is It?	141

7.8   Why Does MaaS Make a Good Departure Point
        for a Smart City?   144

7.9   MaaS Analytics and Their Practical Application   144

7.10   Traffic Management—What Is It?   144

7.11   Why Does Traffic Management Make a Good
        Departure Point for a Smart City?   146

7.12   Traffic Management Analytics and Their Practical
        Application   146

7.13   Transit Management—What Is It?   146

7.14   Why Does Transit Management Make a Good
        Departure Point for a Smart City?   148

7.15   Transit Management Analytics and Their Practical
        Application   149

7.16   Performance Management—What Is It?   149

7.17   Why Does Performance Management Make a Good
        Departure Point for a Smart City?   151

7.18   Performance Management Analytics and Their
        Practical Application   151

7.19   Summary   152

        References   154

8   Transportation Use Cases   155

8.1   Informational Objectives of This Chapter   155

8.2   Chapter Word Cloud   155

8.3   Introduction   156

8.4   What Is a Use Case?   157

8.5   Smart City Transportation Use Case Examples   158

8.6   Summary   160

        References   161

Appendix A: Smart City Transportation Use Case
Examples   162

Use Case Example 1: Asset and Maintenance
Management   162
Use Case Example 2: Connected Vehicle Probe Data   162
Use Case Example 3: Connected, Involved Citizens   163
Use Case Example 4: Variable Tolling   164
Use Case Example 5: Ticketing Strategy and Payment
Channel Evaluation   164
Use Case Example 6: Intelligent Sensor–Based
Infrastructure   165
Use Case Example 7: ICT Management   166
Use Case Example 8: Electric Fleet Management   166
Use Case Example 9: Mobility Hub   167
Use Case Example 10: Partnership Management   168
Use Case Example 11: Transportation Governance
System   168
Use Case Example 12: Customer Satisfaction and
Travel Response   169
Use Case Example 13: Travel Value Analysis   170
Use Case Example 14: Accessibility Index   170
Use Case Example 15: Urban Automation Analysis   171
Use Case Example 16: Freight Performance
Management   171
Use Case Example 17: MaaS   172

9   Building a Data Lake   173

9.1   Informational Objectives   173

9.2   Chapter Word Cloud   173

9.3   Introduction   174

9.4   Definition of a Data Lake   177

9.5   How a Data Lake Works   178

9.6   Value of a Data Lake   182

9.7   Challenges   185

9.8   An Approach to Building a Data Lake   187

9.9	Organizing for Success	190
9.10	Summary	193
	References	193

10 Practical Applications and Concepts for Transportation Data Analytics **195**

10.1	Learning Objectives	195
10.2	Chapter Word Cloud	196
10.3	Introduction	196
10.4	Concepts	197
10.5	Freeway Speed Variability Analysis	198
10.6	Smart City Accessibility Index	214
10.7	Arterial Performance Management	216
10.8	Decision Support for Bus Acquisition	216
10.9	Thoughts on the Use of Analytics	217
	References	220

11 Benefit and Cost Estimation For Smart City Transportation Services **223**

11.1	Informational Objectives	223
11.2	Chapter Word Cloud	223
11.3	Introduction	224
11.4	Overview of the Approach	225
11.5	Assumptions	232
11.6	Smart City Cost and Benefit Estimation	235
11.7	Assumed Configurations for Cost Estimation Purposes	237
11.8	Cost Estimates for Smart City Transportation Services	246

11.9	Smart City Transportation Service Cost Summary	259
11.10	Estimated Benefits for Smart City Transportation Services	259
11.11	Smart City Transportation Services Cost and Benefits Summary	267
11.12	Summary	268
	References	269
12	**Summary**	**271**
12.1	Instructional Objectives	271
12.2	Chapter Word Cloud	271
12.3	Introduction	272
12.4	Review of Chapter 1	273
12.5	Review of Chapter 2	273
12.6	Review of Chapter 3	274
12.7	Review of Chapter 4	274
12.8	Review of Chapter 5	275
12.9	Review of Chapter 6	275
12.10	Review of Chapter 7	276
12.11	Review of Chapter 8	276
12.12	Review of Chapter 9	277
12.13	Review of Chapter 10	277
12.14	Review of Chapter 11	278
12.15	Advice for Smart City Transportation Professionals	278
12.16	Conclusion	281

12.17 Further Reading   282

References   282

About the Author   **283**

Index   **285**

Preface

In developing this book, the author has drawn extensively on his Scottish heritage. In Scotland, there is a long tradition of heavy engineering including shipbuilding and bridge building. The term Clyde-built became a synonym for high-quality engineering that combines innovation with a practical, robust approach. It is hoped that this book will follow in that tradition by offering a robust yet innovative engineered structure that bridges the gap between data science and transportation. It examines the current trend for smart cities, focusing on the type of transportation services that would be required in the smart city. While smart cities are obviously much more than transportation, a focus has been placed on this subject within the smart city context because of the relative importance of transportation in providing the conductivity and accessibility required for a smart city.

Connected and autonomous vehicles have also been given emphasis in this book because of the possibilities for improved efficiency, safety, and user experience that can be provided by driverless vehicles. Connected and autonomous vehicles also have a significant impact on big data and analytics because of the size of the data set that will be generated by such vehicles. This has the potential to revolutionize data collection and analysis for transportation.

The author has had the privilege of working closely with experts in transportation, data science, and analytics in the course of consulting assignments over several years. This has offered a unique insight into the revolution that is taking place in data management and analytics. The concept of big data has emerged, enabling data sets that were previously considered to be too large for storage and manipulation to be treated as a single repository that can yield new insights and understanding. The author is particularly indebted to the people at Teradata Inc. who, in the course of consulting assignments, have provided knowledge, expertise, and experience in data science and analytics that the author has been able to fuse with experience, expertise, and advanced transportation technologies. Only a world leader in big data and analytics could have provided such experience. The opportunity to be a participant in projects that bridge data science and transportation was pivotal in providing the information and background required to create this book. The author believes that there is much to be gained in the application of big data and analytics techniques to smart cities, and hopes that this book will be instrumental in unlocking the power of such technologies and approaches for the good of transportation.

A large number of people from both the data science and transportation industries have provided impact and influence in the creation of this book. In fact, that are too many to mention individually. There are, however, a few people who should receive particular mention because of the significance of the impact on the author's understanding of the subject. The foremost of these is Peeter Kivestu, industry consultant with Teradata Inc. The author has worked closely with Peeter over the past 2 years, learning a lot about data analytics and data science. Over the course of this experience, Peeter has become a good friend and a wonderful source of knowledge and inspiration. Other notable influences include the following people. Thanks to you all for the knowledge and wisdom that you have imparted.

Al Stern, *Citizant*
Eric Hill, *MetroPlan Orlando*
Mary Gros, *Teradata*
Albert Yee, *Emergent Technologies*
Eva Pan, *LA County Metropolitan Transportation Authority*
Matt Burt, *Volpe Center,*
Alberto Belt, *Teradata*
Greg McDermott, *Teradata*
Mena Lockwood, *Virginia Department of Transportation*
Alex Estrella, *SANDAG*
Gregory Kanevsky, *Teradata*
Michael Bolton, *Pace Bus, Chicago*
Alfredo Escriba, *Kapsch,*
Jacek Becla, *Teradata*
Mickey Schwee, *Tableau Software*
Anita Vandervalk-Ostrander, *Cambridge Systematics*
James Dreisbach Towle, *SANDAG*
Mike Riordan, *Teradata*
Apurva Desai, *Kyra Solutions*

James Durand, *Ohio State University*
Mike Smith, *Teradata*
Armand Ciccarelli, *Appian Strategies*
James Garner, *PACE Bus, Chicago*
Mohammed Hadi, *Florida International University*
Ashley Holmes, *Indiana Toll Road*
Jane White, *LA County*
Nicola Liquori, *SunRail*
Bill Malkes, *GRIDSMART*
Jason Trego, *Amazon*
Nikola Ivanov, *University of Maryland*
Bill Thorp, *Kyra Solutions*
Jeff Siegel, *HNTB*
Paul Huibers, *Teradata*
Brandon Lucado, *Teradata*
Jeremy Dilmore, *Florida Department of Transportation District 5*
Peter Thompson, *SANDAG*
Bruce Teeters, *Teradata*
Jim Clark, *Rhythm Engineering*
Petros Xanthopoulos, *Stetson University*
Carrie Feord, *Teradata*
Jim Durand, *Ohio State University*
Piyush Patel, *Kyra Solutions*
Charles Ramdatt, *City of Orlando*
Jim Misener, *Qualcomm*
Prof. Kan Chen, Cheryl Wiebe, *Teradata*
Jim Wright, Ram Prasad, *Kyra solutions*
Chris Bax, *Cubic Transportation Systems*
John Thuma, *ActionIQ*
Randy Cole, *Ohio Turnpike*
Chris Davis, *MIT Enterprise Forum of Cambridge*
Kathleen Frankle, *University of Maryland*
Ray Traynor, *SANDAG*
Chris Francis, *Citizant*
Katie Eastburn, *Kyra Solutions*
Rebekah Hammond Dorworth, *Kyra Solutions*
Chris Kane, *Teradata*
Katie Steckman, *Amazon*
Rick Harmison, *Teradata*
Chris Sullivan, *Teradata*
Kevin Borras, *H3BM*
Rick Schuman, *INRIX*

Cindy Wiley, *Teradata*
Kevin Hoeflich, *HNTB*
Rob Hubbard, *Cisco*
Clay Packard, *Atkins*
Klaus Banse, *ITS Columbia*
Robert Bruckner, *RMB Consultive*
Damian Black, *SQLStream*
Kris Milster, *Traffic Technology Services*
Robert Murphy, *aecom.com*
Dan Graham, *Teradata*
Kyle Connor, *Cisco*
Stephen Payne, *AECOM*
Danielle Stanislaus, *Emergent Technologies*
Luis Hill, *H3BM*
Steve Gota, *LA County Metropolitan Transportation Authority*
Darla Marburger, *Claraview,*
Marc Chernoff, *Teradata*
Tim Lomax, *Texas Transportation Institute*
David Lawson, *NewSci*
Marc Grosse, *Teradata*
Tip Franklin, *Smart Cities Consultant*
Devang Patel, *Kyra Solutions*
Marcie Selhorst, *Teradata*
TJ Ross, *Pace Bus Chicago*
Diane Gutierrez-Scaccetti, *Florida's Turnpike Enterprise*
Mark Demidovich, *Georgia DOT*
Tony Stryker, *Teradata*
Dick Kane, Mark Hallenback, *University of Washington, TRAC*
Tushar Patel, *Florida Department of Transportation District 5*
Dr Mohamed Abdel-Aty, *University of Central Florida*
Martha Morecock-Eddy, *HNTB*
Vatsal Patel, *Litmus Solutions*
Dr. Essam Radwan, *University of Central Florida*
Martin Rosell, *WirelessCar*
Vik Bhide, *City of Tampa*
Ellison Alegre, *SANDAG*
Mary Eward, *Teradata*

1

Introduction

1.1 Introduction

This book explores the boundaries between several major subject areas including transportation, data science, automotive design, and smart city planning. The intention is to provide comprehensive guidance on the major aspects of smart cities from a transportation perspective. This chapter sets the scene for the remainder of the book, orienting readers on the book's content and the level of detail that is provided.

1.2 Informational Objectives of This Chapter

This chapter answers the following questions:

- What is the background within which this book was written?
- Why were the selected subjects chosen?
- Why is the book relevant now?
- Who is the intended readership of the book?
- What topics are addressed in the book?
- What should different readership groups expect to achieve by reading the book?

1.3   Chapter Word Cloud

A good way to provide a visual overview of the contents of a document is by creating a *word cloud*. A word cloud presents the words used most often within a document with the size of the font in proportion to the frequency of use of the word, providing a quick and simple way to overview the document's contents. Figure 1.1 shows the word cloud for Chapter 1. Similar word clouds are provided at the beginning of each of the remaining chapters.

1.4   Background

This is a wonderful time to be a transportation professional. The opportunities that technology and data availability provide are significant, and it is a privilege to have the opportunity to write this book to communicate some of these wonderful things. In 1997, I wrote a book on the use of system architecture techniques in intelligent transportation systems. The motivation to write the book was to make valuable techniques and approaches accessible to a wide range of transportation professionals and others interested in applying system engineering and systematic data analysis techniques to transportation. That book focused on the interface between system engineering and transportation. This book presents a similar opportunity with potentially more powerful results. This new opportunity lies in addressing the interface between transportation and data science, and thereby illuminating the usefulness and power of recent advances in data science and analytics within the context of smart cities and transportation. This book is driven by the same motivation as that for my ear-

Figure 1.1   Word cloud for Chapter 1.

lier work: to bring to the attention of transportation professionals the existence of powerful possibilities for realizing the value of data in terms of safety, efficiency, and enhanced user experience.

Much of the book's content is derived from a series of consulting engagements conducted by my company on behalf of Teradata, Inc. Teradata is the leader in big data and analytics in multiple markets including airlines, banking, and retail. The consulting engagements were designed to support the introduction of this experience and expertise into the worlds of intelligent transportation systems and smart cities. These assignments afforded a unique opportunity to explore and define the boundary between transportation and data science, in cooperation with leading big data and analytics experts and in coordination with transportation clients.

This book appears as we experience the convergence of rapidly rising data science capabilities and the growing availability of a wide range of transportation data. In fact, there is probably more data available regarding transportation infrastructure and operations than at any time in the history of transportation. Initiatives for smart cities, connected vehicles, and autonomous vehicles promise to add even more volume to the data already available from infrastructure-based sensors.

Smart cities is an umbrella term that has been widely adopted to address the application of advanced technologies to enhance service delivery and to improve the lives of both citizens of and visitors to cities. According to the United Nations, more people now live in cities than in rural areas [1], raising the importance of smart cities. The connected vehicle delivers a two-way communication ability between the vehicle and the back office. This book uses the term *back office* to refer to an off-road processing or management center that receives data from vehicles and roadside infrastructure and subjects it to data processing that will convert it into information, insight, and understanding. The term back office is not typically used in transportation; however, the book describes several different processing and management centers, and it is useful to have a single general term. The two-way communication capability of a connected vehicle also enables vehicles to communicate with each other, offering some significant safety improvements through the avoidance of potential crashes. The vehicle-to-back office connectivity also allows for the extraction of a large volume of data from vehicles and the delivery of information to drivers. Autonomous or self-driving vehicles make possible private cars that relieve drivers of the burden of driving and freight vehicles and transit vehicles that no longer require a driver. These autonomous vehicle developments require significant amounts of data for management and control purposes; at the same time, these advanced vehicles will generate substantial amounts of data.

Private sources of data have also emerged in recent years adding to the extreme volume of data that is now accessible to transportation professionals and others looking to understand the trends and patterns associated with transportation. Transportation professionals are endeavoring to extract value from the new data by managing it and converting it into information. This book aims to recognize and address the new challenges and opportunities faced by transportation professionals by providing input to decision-making and strategy development associated with the application of big data and analytics to transportation service delivery within a smart city.

1.5   Why This Subject and Why Now?

The thinking behind this book is based on a simple question: How can all the valuable lessons learned about big data and data science be assembled in a single source? Further, how can I present these lessons in a way that is meaningful for people who are interested in applying these techniques to transportation in smart cities? I have been fortunate to be able to focus over the past two years specifically on the application of big data and analytics techniques to transportation. Building on a background in transportation, this period has been full of revelations and new insights into how emerging data science and analytics technologies may benefit transportation in smart cities. A vantage point from the boundary of transportation and data science offers a unique perspective on how the two subject areas can interrelate. Therefore, a definition of success for this book would be to build awareness and interest within the smart city community, motivating practitioners to become familiar with and to make use of the new tools and capabilities available. Significant power can be released through the effective use of big data and analytics for smart city transportation. While the subject may have become somewhat hackneyed due to the widespread use of the terms big data and analytics, there are significant possibilities in applying these principles to smart city transportation. The fact that the terms are popular can also be considered positive since it implies that a wider group of people has at least an awareness that something is going on. Unfortunately, this awareness is not always based on a firm foundation. This is another reason for writing this book at this particular time. The hope is that this general interest in big data and analytics will serve as a platform for adding additional information and shedding light on the true nature of transportation mechanisms for both supply and demand.

The nature of the revolution that is taking place in data science is not just related to the sheer size or volume of the data. Exciting possibilities for transportation exist in the fact that as the data sets get bigger, they also become richer in insight and understanding. So it's about having the capability to store and manipulate vast amounts of data *and* obtaining better information from which deeper insights can be obtained. This is enabled by a wider data horizon that stretches across the entire enterprise or organization. This also offers the ability to merge data into a centrally accessible repository, whereas previously it may have been stored in a stovepipe fashion to fit within the constraints of prior data management technologies. This new ability to obtain an enterprise-wide view of data multiplies the information possibilities by a large factor.

1.6 Intended Readership Groups for the Book

Defining the intended readership for the book, serves two purposes. First, it provides the author with a clear understanding of the intended target readership to guide the selection and presentation of the content. Second, it enables readers to understand the value that can be realized by reading the book. The book intends to address the following readers:

- Transportation planners and traffic engineers;
- Business analysts, data scientists, data engineers, and developers;
- Automotive manufacturers;
- Smart city advocates and implementers;
- University professors and students.

The following sections describe each of these groups and explain the value they can obtain from reading the book.

Transportation professionals.

This group includes transportation planners, traffic engineers, transportation operations and service delivery specialists, freight planners and freight operations professionals, and executive leaders in transportation. The book aims to provide this group with an understanding of what big data and analytics are all about and how they can be applied in a practical and useful way to transportation planning, operations, and service delivery.

Business analysts, data scientists, data engineers, and developers.

Analysts, scientists, engineers, and developers are likely to be involved in developing big data and analytics systems or in applying commercial off-the-shelf solutions to transportation problems. The book provides these readers with an overview of how their technologies and solutions can be applied in a practical and useful way in the transportation domain.

Automotive manufacturers.

This group includes engineers, product planners, and designers working within automotive manufacturers. It also includes equipment and service providers that work with automotive manufacturers to deliver connected and autonomous vehicle solutions to drivers. The book aims to provide this group with a thorough understanding of how big data and analytics can be applied to transportation in a practical way. The value of connected and autonomous vehicles within the bigger picture of data supply, data conversion to information, and the use of analytics to understand trends and patterns are also discussed. Automotive manufacturers and their electronics suppliers are deeply involved in both connected and autonomous vehicles. This book should provide valuable input into ongoing decision-making and product development.

Smart Cities planners, advocates, and implementers.

This readership group includes city officials involved in planning the transportation elements of a smart city, city mayors, and other executive leaders charged with implementing a smart city vision. The book provides this group with information on the elements that can be incorporated into a smart city from an urban analytics perspective. Further, it illustrates the power of big data and provides a solid business justification for incorporating big data and analytics into any smart city vision. This is an exciting time for city transportation with a strong focus on developing smarter and better cities. The assumption is that big data and analytics will form a central core for advances in smart city transportation, founded on a detailed understanding of transportation supply, demand, and operating conditions. This understanding is expected to go a long way toward improving service delivery for transportation in a city.

University professors and students.

For this readership group, the book aims to provide an up-to-date and useful resource on the world of big data and analytics within a smart city context. It is hoped that an explanation of how the latest techniques and technologies can be applied to transportation will spur the next generation to use current efforts as a platform and move toward establishing the role of data and data analytics and transportation. In many respects, big data and analytics are building blocks that represent the new asphalt, concrete, and steel.

1.7   Overview of Contents

Subsequent chapters of this book address a range of subjects related to understanding big data and analytics, showing their relevance to transportation, and then defining what can be done.

Each chapter begins with a list of informational objectives for the chapter. This is an application of instructional system design intended to ensure that the purpose for each chapter is clearly defined in advance so that it can be addressed in the ensuing content. As I consider instructional system design to be the application of system engineering techniques to instructional materials, I thought it would be appropriate to apply this technique in this book.

The content of Chapters 2–12 is summarized in the following sections.

Chapter 2

This chapter provides an overview of the questions that can be addressed by big data and analytics tools and techniques. The traditional system engineering approach to the development of a solution or answer to a problem defines the problem with a set of requirements. To a system engineer, this seems logical. However, experience has shown that that when people describe an issue to be addressed and subsequently understand possible solutions, they change their minds about the problem based on the new information that they have just received about what can be done. For this reason, the book addresses early on the definition of questions that can be answered through the application of big data and analytics techniques. The idea is to fully support the operation of a what-how cycle within the context of the book, through the adoption of a specific methodology as shown in Figure 1.2

The what-how methodology begins with questions to be addressed, which feed into requirements that incorporate needs, issues, problems, and objectives, defining the questions to be addressed, explaining what big data is, and then defining what can be done. Using a combination of the questions to be addressed and the definition of the solutions that come later, it is possible to enable readers to develop their own customized methodology to ensure that final solutions take full account of what needs to be done and how it can be done. Defining the questions to be addressed is placed at the beginning of the methodology to frame the subject and to introduce readers to the value of big data and analytics.

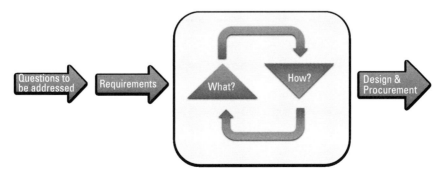

Figure 1.2   The what-how cycle methodology.

This is done in a practical way by defining questions that are valuable and relevant to transportation today. In some cases, readers may already be aware of the questions and issues that are defined here; it is hoped, however, that the descriptions given will reinforce readers' preexisting knowledge and lend a better grasp of the practical application of big data and analytics.

The questions to be addressed can be considered as essential elements within the bridge between data science and transportation. It can be assumed that the questions lie firmly on the transportation side of the bridge and that the solutions lie on the data science side. Through defining the questions, we begin the construction of the bridge and get our readers ready to walk across it

Chapter 3

This chapter explores the nature of big data with the intention of providing a solid overview and understanding of this topic. This is not a data scientist's definition of big data; rather, it is a transportation view of data science subjects. Chapter 3 addresses the different dimensions of big data, the importance of big data, and the relevance of big data to transportation in a smart city. In addition, Chapter 3 provides examples of big data sources that exist within the transportation ecosphere and the nature of that big data. The overall objective is to thoroughly define the nature of big data and its role in forming the enabling platform for analytics.

Chapter 4

Chapter 4 discusses connected and autonomous vehicles. These are two different subjects that are related using advanced technology for vehicles, sometimes referred to as telematics. Big data and analytics have a significant role to play in the connected vehicle. The connected vehicle involves the concept of linking vehicles to roadside infrastructure and vehicles to other vehicles with wireless technology. Essentially this capability enables data to be obtained from the vehicle and information to be provided to the driver while the vehicle is in motion. Chapter 4 is an important element of the book because of the huge potential data source that connected vehicles represent.

A recent article in *Forbes* magazine [3] featured the Ford Fusion Energi plug-in hybrid. This car, which achieves 108 miles per gallon, generates 25 GB of data every hour. Extrapolating this data rate across the entire U.S. vehicle fleet, then connected vehicles could generate approximately 2 ZB of data every year. This is a difficult number to comprehend, even by data scientists, but consider that in 2013, the entire World Wide Web generated 4 ZB. This offers some idea of the potential scale of connected vehicle data—50% of the entire World Wide Web's volume. Admittedly, this estimate is likely to be on the high end because not all vehicles will be as smart as this particular Ford vehicle. However, it does give a sense of the impact that big data from connected vehicles will have on transportation. The big data and analytics aspects of connected vehicles include managing the volume of data coming from the vehicle, conducting analytics in the back office, and then projecting those analytics and results back to drivers and other people associated with traffic and transportation operations and management. Chapter 4 also explains autonomous vehicles, with feature the ability to operate without a driver.

Chapter 5

Chapter 5 discusses the concept of a smart city. Smart City initiatives have commenced in many cities around the world. With respect to the United States, a recent smart city challenge issued by the U.S. Department of Transportation [2] attracted 78 responses from U.S. cities, with the city of Columbus, Ohio, selected as the winning applicant. It is obvious that smart cities go beyond transportation and incorporate social services, social networks, energy grids, and smart places to live and work. Chapter 5 aims to take a close look at smart cities from a transportation perspective, especially as one of the core features of the smart city will be extensive use of big data and analytics. Chapter 5 also briefly outlines other services that could be expected in a smart city, showing how big data and analytics along with transportation services fit into the bigger picture.

Chapter 6

Chapter 6 explains data analytics, with particular attention to its transportation aspects. Good analytics go beyond even world-class reporting and provide transportation professionals with the opportunity to influence the performance of their organizations. Accordingly, Chapter 6 discusses the nature of data analytics and presents a few examples from outside of transportation to afford readers a solid understanding of this topic.

Chapter 7

Chapter 7 builds on Chapter 6 by detailing information regarding analytics that can be applied to transportation. The intention is to bring the term analytic to life from a transportation perspective. Accordingly, Chapter 7 includes descriptions of analytics that have been applied and analytics that could be applied in the future. These are analytics that can be directly related to safety, efficiency, or user experience improvements. Chapter 7 also discusses the relevance of these applications to different aspects of smart city transportation such as planning, operations, and maintenance.

Chapter 8

Chapter 8 adopts the system engineering term *use cases* to capture systematically what can be done with big data and analytics. In addition, Chapter 8 presents a catalog of transportation use cases that are relevant to a smart city. These will be presented in a standard format that contains the following use case attributes:

- Smart city transportation service addressed;
- Use case name;
- Objectives;
- Expected outcome of analyses;
- Success criteria;
- Source data examples;
- Business benefits;
- Challenges;
- Analytics that can be applied.

This chapter will provide the reader with an explanation of how the tools and techniques contained in the book can be applied to smart city transportation services. It also provides a logical structure that can form the basis for a use case catalog for a smart city. A catalog is not intended to be a comprehensive prescription but is a model on which to base thinking across the smart city. This is designed to enable progress toward formalizing descriptions of what big data and analytics will do. This is part of the what-how cycle, introduced in this chapter. This progress involves an evolution of the knowledge of what is required to address practical problems, in light of a growing understanding of how it can be done.

Chapter 9

The concept of a *data lake* is used to communicate the creation of a central, accessible, discoverable body of data. Chapter 9 discusses and explains a robust approach to the creation of a data lake. The approach incorporates the pitfalls and challenges encountered in previous projects with the creation of a data lake and the lessons learned from those difficulties. It is not intended to be a one-size-fits-all recipe for the creation of a data lake, but rather, a model from which to build customized approaches for each implementation.

Chapter 10

Examples of the implementation of the techniques and concepts contained in this book provide a powerful tool for explaining the relevance and usefulness of the book's contents. Ideally, the examples would represent a full implementation of the techniques and concepts. However, since the application of big data and analytics to transportation is still in an early stage, there are not too many implementations that represent full-fledged applications. Therefore, Chapter 10 presents a combination of implementations and concepts. Some of the examples represent implementations that have been conducted with clients, while others represent project concepts that have been discussed with clients but not yet implemented.

Chapter 11

Chapter 11 presents a methodology for estimating costs and benefits for all 16 of the smart city transportation services defined in Chapter 5. Several factors, including the size of the data, the number of users, the types of queries to be supported, and the query speed to be achieved, affect costs. Thus, cost estimates require the kind of detailed system design that is beyond the scope of this book. However, it is possible to provide a set of yardsticks and expected cost ranges based on past experience. This enables the definition of budgetary costs for incorporation into future plans and work programs. Accordingly, Chapter 11 defines and explains model configurations as the basis for cost estimates.

In addition, Chapter 11 explores the values and benefits of applying big data and analytics techniques to transportation. Starting with the core values of safety, efficiency, and enhanced user experience, it is possible to extrapolate a range of values and benefits that can be achieved through the application of big data and data analytics. The exact benefits and values to be derived will be significantly influenced by the nature of the problem and by the details of the selected design. However, it is possible to provide some ranges and estimates for the values and benefits that can be expected. These should be of value in creating a business justification for big data and analytics investments. Chapter 11's cost estimates and benefits assessment concludes with the development of benefit cost estimates. While these are approximate, they assist in defining a sketch planning approach to smart city transportation service evolution.

Chapter 12

The book concludes with Chapter 12, which provides an overview of the essential elements covered in the book. Specifically, Chapter 12 discusses the relevance of big data and analytics with respect to transportation, smart cities, and the connected vehicle. In addition, Chapter 12 draws conclusions on the value of big data and analytics to transportation and the steps that should be taken toward harnessing these powerful resources. Finally, Chapter 12 distills the essential information provided in the book to define advice for smart city professionals, providing a concise summary of actions to consider after digesting the contents of the book.

References

[1]   United Nation, Department of Economic and Social Affairs, Population Division (2014), *World Urbanization Prospects: The 2014 Revision, Hightlights (ST/ESA/SER.A/352)*.

[2]   U.S. Department of Transportation, Notice of Funding Opportunity Number DT-FH6116RA00002, "Beyond Traffic: The Smart City Challenge—,Phase 2" March 21, 2016, https://www.transportation.gov/smartcity/nofo-phase-2, retrieved April 3, 2017.

[3]   McCue, T. J., "108 MPG with 2013 Ford Fusion Engeri, Pluse 25 Gigabytes of Data," *Forbes,* January 1, 2013, http://www.forbes.com/sites/tjmccue/2013/01/01/108-mpg-with-ford-fusion-energi-plus-25-gigabytes-of-data/#23ac1cd14coa5.

2

Questions to Be Addressed

2.1 Informational Objectives of This Chapter

This chapter addresses the following questions:

- Why are big data and analytics important?
- How can they deliver value?
- Why questions and not answers in this chapter?
- What is the value of data?
- What questions can I answer by applying big data and analytics?
- What can I do with the questions once I have them?

2.2 Chapter Word Cloud

Figure 2.1 presents a word cloud with an overview of the content of this chapter.

2.3 Introduction

In past discussions with clients who are interested in harnessing the power of big data and analytics, the client often asks the following question: "How do we get started?" One simple initial response would be, "Stop throwing the data away." This might seem like a strange place to start, but the fact is that many transpor-

Figure 2.1 Chapter 2 word cloud.

tation agencies, for a variety of reasons, don't retain raw data. The reasons range from misconceptions about the cost and complexity of storing data to concerns about liability associated with keeping the data. Sometimes, another factor is at play, namely a perception issue related to the value of data. Unfortunately, data is not that attractive. A large volume of data is not compelling unless you're a data analyst or data scientist. The real value in data can really only be realized when it is converted into information, related to the users' needs and to the jobs that the user is supporting daily. Asking people to preserve something makes it particularly important to communicate the value of it. The challenge is to define a way to alter the perception of data as a raw material. Consider clay, a relatively common raw material that is used extensively in construction. Clay is a somewhat unattractive substance found in the ground and has seemingly little value. It often has an offensive odor, and it is hard to perceive that this material could be the basis for future value.

In London, many of the buildings are built from a famous type of brick made from yellow clay. These buildings, in many cases, carry an extremely high perceived value and yet began their life as humble clay. It seems that if thought can be directed forward across the whole value chain from the basic material through the various conversion processes to the ultimate product that delivers value, then the perception of the original basic material can be altered.

So what does clay have to do with big data and analytics? Just like clay, our perception of the value of raw data is low. Some transportation agencies even summarize data and then throw the raw data away, while others don't even bother collecting data. If, however, the transportation industry were to develop a future vision along the process that converts data to information, insights, and actionable strategies, then perhaps it would begin to recognize data as a raw material that could transform transportation in smart cities. Many believe that the main issue is the inability to store large volumes of data because of cost constraints. Fortunately, our new perception of the value of data coincides with new abilities to store huge volumes of data at relatively low cost. This issue is addressed in Chapter 11, which covers cost and benefits estimates.

Advances in data science enable the storage and manipulation of data in ways that were not thought possible in the relatively near-term past. Thanks to Hadoop, Google, and Amazon, some important new abilities to store and manipulate data are converging with the new perception of the value of data. There are also new possibilities for structuring and restructuring data to optimize its retrieval and management.

For example, a new concept known as a data lake allows data from numerous sources to be merged into a single repository. Data lakes—essentially centralized data repositories that can ingest data from multiple sources and make it accessible across an organization or enterprise—can contain a wide variety of structured and unstructured data. In data lakes, the data is clean, contiguous, and easily accessible—unlike the contents of data swamps. Thus, data lakes enable an enterprise-wide view of data. Just like magnificent brick buildings wrought from humble clay, the data lakes and integrated data exchanges of the future can be constructed from data as their raw material. Chapter 10 further details data lakes.

2.4 Questions Instead of Answers

This chapter addresses a critical ingredient of the book. To engage the reader and explain why it's worthwhile to understand big data and data analytics, it is important to define and frame the questions that can be addressed by such tools and techniques. The chapter also provides a high-level framework that paves the way for the discussion of use cases in Chapter 9. The questions to be addressed encapsulate the needs, issues, problems, and objectives that transportation professionals encounter. Identifying and defining the questions to be addressed is also a first important step in a results-driven approach to the use of big data and data analytics. An understanding of these questions will form the basis for the definition of use cases and objectives for the analysis work.

2.5 Overview of the Questions

The questions contained in this chapter are not designed to be completely exhaustive but rather, to show some examples of the kinds of questions that can be addressed by the application of big data and analytics techniques to transportation. The intent is to confirm what can be done and raise awareness within the transportation profession on the sorts of questions that can be asked and that are probably already being asked in the wider world of commerce and business beyond transportation. There is a wide variety of categories to describe performance measurement and management for transportation; performance measures are typically supplied as answers to the questions. An overview of these performance categories reveals that they can all be traced back to three root performance areas: improved safety, increased efficiency, and enhanced user experience. (This assumes that the environmental effects of transportation can be categorized under the efficiency performance area.) These three categories will be used to structure the list of questions that can be answered through the application of analytics, summarized as follows:

- Safety-related questions;
- Efficiency-related questions;
- Enhanced user experience-related questions.

Based on conversations with transportation professionals regarding their needs, issues, problems, and objectives, I have created a list of 20 big questions. The term big questions is used to indicate that these are high-level questions to which several more detailed questions could be associated. It is likely that some of the big questions will be more pertinent to different readership groups than others. To address this variation, Table 2.1 relates the big questions to their likely readership groups.

The reader may have a question at this point in the book: why are we discussing questions instead of the answers? The answers will be provided as we move through the book. However, since the adoption of big data and analytics techniques in smart city transportation is at an early stage, there is considerable value in framing the questions as well as providing the answers. Another important reason for defining the questions is that this is a positive way to explain the needs, issues, problems, challenges, and objectives that can be addressed by big data and analytics. A question can also be thought of as a focused starting point, the initial step on the road toward big data and analytics. In many implementations beyond transportation, the quest for big data and analytics has started with a question or a problem to be addressed. Ultimately, the systems implemented are capable of extremely flexible and varied application right across a business or enterprise. Experience in the application of advanced technologies to transportation suggest that taking a small focused step that delivers immediate results while providing the basis for business justification and taking the next step is always the best way to ensure a successful application of

Table 2.1
The 20 Big Questions

Questions	Transportation Planning and Traffic Professionals	Transit Planners and Operating Professionals	Freight Planners and Operating Professionals	Business Analysts, Data Scientists, Data Engineers, Developers	Automotive Manufacturers	Smart Cities Advocates and Implementers	University Professors and Students
Safety							
1 How do we maximize the safety of the transportation system?	●	●	●	●	●	●	●
2 What are the effects of safety improvements?	●	●	●	●	●	●	●
Efficiency							
3 What are the bottlenecks and slowdowns and the transportation system?	●	●	●	●	●	●	●
4 How do we optimize the efficiency of the transportation system?	●	●	●	●		●	●
5 How are transportation assets performing, and how can they be managed better?	●	●	●	●		●	●
6 How do we optimize current and future expenditures on operations and capital?	●	●	●	●		●	●
7 What impact will connected and autonomous vehicles have on the transportation system?	●	●	●	●	●	●	●
8 How do we make it easier for transportation customers to pay?	●	●	●	●		●	●
9 How do we improve service levels for citizens?	●	●	●	●		●	●
10 How do we improve service levels for visitors?	●	●	●	●		●	●
11 How do we optimize land-use?	●	●	●	●		●	●
12 What are the service deficiencies in the transportation system?	●	●	●	●		●	●
13 What is the current demand, and what will be the future demand for transportation?	●	●	●	●		●	●
14 How do we maximize access to jobs?	●	●	●	●		●	●
User Experience							
15 How do we enhance the user experience for travelers in the transportation system?	●	●	●	●	●	●	●
16 Are transportation customers getting the best value for money?	●	●	●	●		●	●
17 How do transportation customers perceive service levels?	●	●	●	●	●	●	●
18 What impact will mobility as a service have on the transportation system?	●	●	●	●	●	●	●
19 How can travelers make the best use of the transportation system?	●	●	●	●		●	●
20 How will better information change travel behavior?	●	●	●	●	●	●	

the technology. That's why the concept of the project is so important within transportation because it's focused, it has objectives, it has structure, and it has a clearly defined set of questions to be addressed. Projects can also be scheduled and budgeted accurately. It is not a question of avoiding the identification and definition of the answers and solutions. That will indeed come in Chapters 6 and 9, which discuss how to build a data lake and the nature of analytics. As a starting point, Sections 2.6–2.8 explore the questions that can be addressed.

Defining the questions also begins the thought process on another important aspect of the organizational framework that will support big data and analytics within the organization. Within private sector organizations, there is typically one or more business analyst focused on the use of data and the extraction of information to improve the performance of the business. Typically, this type of role is less prominent within transportation organizations, as their primary focus is on planning, designing, delivering, maintaining, and operating transportation services. While information technology plays a significant role in the modern transportation world, the use of data and the extraction of value from data is still an emerging subject. Accordingly, it is hoped that these questions will also start the discussion within transportation enterprises on who will be responsible for big data and analytics and how this role will fit within the overall context of the organization. In addition, questions assist us in defining what we are trying to achieve. Furthermore, questions can have a powerful effect by focusing thought and subsequent actions. For example, when my discussions with one client had initially focused on the virtues of big data and analytics, I suggested that a transportation data lake be created and that transportation data analytics capabilities be acquired to enable data discovery. After several minutes, the client asked the following question: "Can you relate all this terminology to performance management or active arterial management because these subjects are addressed in the current work program and are eligible for funding?" This not only illustrates the power of a question, it also highlights the fact that concepts, no matter how good they are, must fit within the current needs of the organization and, moreover, should align with current organizational momentum. Given that procurement cycles and planning for the implementation of new technologies are relatively long in the public sector, it is a good idea for transportation agencies to continue with ongoing initiatives. These can be supplemented and enhanced to achieve the same effect as a radical departure. Another purpose of questions is to help transportation agencies to identify a suitable starting point that relates to current challenges. The starting point must also fit exactly with organizational momentum regarding preexisting plans. In other words, they need to tackle previously identified business problems for which big data and data analytics solutions have not been considered.

Before beginning the exploration of the 20 big questions, let's consider four questions that are extremely valuable in considering transportation as a system. These are paraphrased from a keynote speech at the Intelligent Transportation Society of America's annual meeting in Houston in 2010 in which the then chairman of IBM, Sam Palmisano, brought IBM's vast experience in developing systems to bear on transportation [1].

In that speech, Palmisano indicated that IBM considers the following four questions when determining if a system is really a system:

- Does it have clarity of purpose?
- Are the major elements connected?
- Can we determine the status at any given time?
- Can it adapt to changes in the environment?

Clarity of purpose refers to a predefined and agreed on set of objectives for the system. Transportation does pretty well in this respect, although we tend to have purposes and objectives for individual elements such as toll roads, transit systems, and freeways, rather than a single set of objectives for an entire transportation network within a city. This leads to the second question regarding connectivity. There has been a tendency to stovepipe transportation, attaining the benefits of specialization and close control. Such stovepiping is ineffective when it comes to applying advanced technologies, as many of those technologies are designed for sharing. For example, if your work requires you to travel across the country, it is not necessary to buy a plane to get there. Anyone can purchase a plane ticket that enables the sharing of a single plane by many people. Many advanced technologies are just like this; it makes much more sense to use various mechanisms to identify people who can share the cost and effort involved in achieving a given objective.

The question about status directly addresses the use of sensors and, ultimately, probe vehicle data for transportation, which makes extensive use of sensors and telecommunications networks to link transportation elements together. It can be expected that the volume of data from these sources will grow extremely rapidly over the next few years.

The final question on adaptability is one of the big reasons why we need to continue to apply advanced technologies to transportation. It is also why progress toward a complete and detailed understanding of current operating conditions and future demand for transportation is important. Many transportation projects involve the use of asphalt, concrete, and steel. Scheduled durations for these projects can be multiyear, with the decisions made now, setting the scene for 30–50 years into the future. There is not a great deal of flexibility in infrastructure. Fortunately, however, the dynamic nature of transportation demand and our ability to manage capacity for better results stands in contrast to this inflexibility concerning our infrastructure.

These four simple, yet powerful questions go a long way to framing the current state of transportation and the actions and investments required to improve service delivery and to manage transportation in a much more effective manner.

Again, the 20 big questions are summarized in Table 2.1. As discussed earlier, the questions are categorized as safety-related, efficiency-related, and user–experience related (with the assumption that environmental effects are included under the efficiency umbrella). While Table 2.1 identifies the intended readership groups for the questions, readers are encouraged to explore all of the questions to gain a complete overview of the subject. Please note that the definition of these high-level questions does not represent a deep dive into each of the subjects. The intention is to waterski across the subject matter to provide more complete coverage at the expense of detail. Sections 2.6–2.8 discuss the questions further.

2.6   Safety-Related Questions

The safety-related questions focus on ways in which safety can be improved through crash reduction and incident management, while considering the cost of improvement.

How Do We Maximize the Safety of the Transportation System?

The safety of a transportation system can be measured in multiple dimensions. For a start, there are different modes of travel such as private car, transit, bicycle, pedestrian, and freight. There are also different dimensions to travel safety. These include crashes, incidents, the deployment of emergency resources, incident response, and, of course, human behavior. Having an accurate picture of the total number of crashes and the type of crashes is just a starting point. Other questions that can be addressed by big data and analytics include those relating to causal factors such as street lighting, road width, traffic speeds, weather conditions, the presence of a sidewalks, and geometric parameters. Using analytics, we can determine the relationship between these causal factors and support the sort of data discovery that will lead to other questions. In the retail business, for example, large companies use big data and analytics to establish the probability that customers who buy product A will also purchase product B. This information can be used to locate products in close proximity to each other and predict demand for one product based on sales of the other. Of course, analytics can only be achieved if a suitable central repository of data (or data lake) has been created, combining all the data regarding the causal factors described above. If data is then added regarding investment programs or work programs for safety improvements, a whole new set of questions regarding the effects of safety improvements can be posited. A simple before-and-after comparison of key factors can identify the precise effects of a safety improvement. It may also be necessary to take account of other factors that may be influencing improvement in safety, such as overall economic conditions and enhancements in vehicle and road technology.

What Are the Effects of Safety Improvements?

Ideally, an investment in a safety improvement will lead to a positive return on investment, with the safety benefits quantified exceeding the overall cost of the improvement. This question has two dimensions. On the one hand, it is necessary to measure the effect of the safety improvement in terms of accident reduction, reduction in accident severity, and the overall impact of crashes on the economy. On the other hand, it is also necessary to measure the investment in terms of capital and operating funding. Estimation of these costs is required to answer the question completely in terms of absolute impact and efficiency of the safety investment in terms of the improvement per dollar.

2.7 Efficiency-Related Questions

The efficiency-related questions are many and varied but all revolve around obtaining the highest value from the smallest amount of expenditure. This also includes the prioritization or targeting of investments to ensure that money is spent wisely. This can also include an assessment of complementary investments that reinforce each other rather than conflict with, or negate, each other.

Where Are the Bottlenecks and Slowdowns in the Transportation System?

This question can be asked differently depending on the mode of transportation being considered. For example, the private car mode question can be addressed by measuring traffic speeds or travel times and comparing these across the entire road network. For a transit system, absolute vehicle speed is less important than station-to-station travel times and the overall time from the traveler's origin to final destination. Bottlenecks can be identified based on a preprepared and agreed on definition or template. This can be used to identify bottlenecks within a large data set using pattern analysis.

From a freight perspective, end-to-end travel times are an important part of the question along with delays incurred in interchange between freight modes—for example, the time taken to move the load from rail to road or vice versa.

How Do I Optimize the Efficiency of My Transportation System?

This is a system-wide question since a transportation system is comprised of multiple modes—private car, transit, freight, airlines, and potentially, ferries. The question has multiple subcomponents, including what is the current efficiency level of each mode of transportation? and what is the current efficiency level of all modes combined? This question would probably be asked in terms of overall origin to destination times for travelers using multiple modes.

How Are Transportation Assets Performing and How Can They Be Managed Better?

Transportation assets consist of infrastructure such as asphalt, concrete, and steel, along with the telecommunications networks and devices used for data collection and control. Examples of the former include roads and bridges, while the latter includes dynamic message signs, fiber-optic networks, and roadside, infrastructure-based sensors such as traffic speed and flow measurement devices. This question highlights the need to have an accurate inventory of infrastructure and devices. The question regarding type and location of existing assets then leads to a further question on desired performance levels for infrastructure and devices.

How Do We Optimize Current and Future Expenditures on Operations and Capital?

A typical approach to defining budgets and work programs for transportation expenditures is to take the amount spent in previous years and add a percentage for future years. While this has been effective, we can do a lot better by asking questions such as, What were the effects of previous investments? What will the effects of my future investments be? This leads to questions such as "To achieve a 1% shift of modal split in favor of transit, what investment should be made and when?"

What Impact Will Connected and Autonomous Vehicles Have on the Transportation System?

The U.S. Department of Transportation estimates that connected vehicles may be able to reduce 70–80% of nonimpaired crashes in the United States [2]. It can be expected that autonomous vehicles will take this even further by completely removing driver risk. Asking this question leads to a lot more questions about what the impact of connected and autonomous vehicles is likely to be on various aspects of transportation. These aspects include fleet management, transit systems, electronic toll collection systems, and the development of mobility as a service (MaaS). MaaS involves providing a portfolio of public and private services as options to travelers. These include things like Uber, taxi services, and transit services, enabling travelers to choose the best option for the trip and the circumstances.

How Do We Make It Easier for Transportation Customers to Pay?

In most cases, transportation customers have to pay a fee in return for service. This can include transit fares, tolls, and parking fees. Often the payment of the fee at the point of service can lead to delays and congestion. The payment of the fee may also be perceived as a barrier to use of the system. This question addresses the ways in which technology can be applied to make it easier for users to pay for transportation services. This leads to other questions such as: can I set up a single system to pay for all transportation modes within my city? Similarly, one might ask can I use the data collected from my citywide transportation payment system as input and for better operational management and planning?

How Do We Improve Service Levels for Citizens?

This question requires us to define what exactly we mean by service and service level and what target service levels we are aspiring to for our cities and our citizens. A service can be considered to be something of value. Table 2.2 summarizes the user services that were identified as part of the National Intelligent Transportation Systems Development program [3], a multiyear effort to define a national framework for the application of advanced technologies to transportation. The program defined eight user service bundles or categories, which are detailed in Tables 2.3–2.9. A brief explanation of each user service is provided in the second column of each of the tables (Tables 2.2–2.9). These notes are not based on the National ITS program work but on my own interpretation.

In transportation, there tends to be a focus placed on the project rather than the service. It could be viewed that projects are a means to delivery of service but that it is the services that deliver the ultimate value in terms of safety, efficiency, and enhanced user experience. A service evolution approach to the application of advanced technologies such as connected and autonomous cars and smart cities could be a very effective way to define a rollout program over time, space, and level of service. With respect to level of service, there is another component to this question. Having identified the services that we intend to deliver to our citizens then what is the desired level of service to be delivered? This will undoubtedly involve other questions like how much are customers prepared to pay?

How Do We Improve Service Levels for Visitors?

It is important to consider visitors to a city as well as its citizens or residents. For many cities the economy is substantially impacted by the activities of visitors as well as residents. The same comments hold good for visitors as the ones defined for citizens above, with the added caveat that visitors may have additional needs related to a lack of understanding of the transportation system and unfamiliarity with the city. This may well lead to additional services being defined and

Table 2.2
National ITS Architecture Program Travel and Traffic Management User Services Bundle

User Service	Notes
1.1 Pretrip travel information	The provision of traveler information services to travelers before they embark on a trip.
1.2 En-route driver information	The provision of traveler information services within the vehicle for drivers of vehicles.
1.3 Route guidance	Turn-by-turn instructions to enable drivers to get efficiently from point A to point B. This can be based on historic data or can incorporate current traffic conditions.
1.4 Ride matching and reservation	Matching demand for ride-sharing with available capacity and providing travelers with prereservation capabilities. Uber is a great example.
1.5 Traveler services information	The delivery of information to travelers regarding services that are available and that may be needed on the journey.
1.6 Traffic control	This includes traffic control services provided by traffic signals and by traffic control and dynamic message signs on freeways.
1.7 Incident management	Services required to manage the incident process from detection through verification and response to clearance.
1.8 Travel demand management	Services to identify and assist in the management of the demand for travel.
1.9 Emissions testing and mitigation	Services to measure, monitor, and mitigate the environmental effects of the transportation process.
1.10 Highway rail intersection	Services to manage the potential conflict between rail vehicles and road vehicles at grade railroad crossings.

Table 2.3
National ITS Architecture Program Public Transportation Management User Services Bundle

User Service	Notes
2.1 Public transportation management	Services to automate the operations planning and management of public transit systems, including buses and railed vehicles.
2.2 En-route transit information	Services to provide real-time information to transit system passengers.
2.3 Personalized public transit	Services that provide demand-actuated transit services to individual travelers.
2.4 Public travel security	Services that protect and enhance the security of the traveling public.

delivered specifically for visitors, and this would of course require additional questions.

How Do We Optimize Land Use?

This rather broad question has several related subquestions. The first of these would be, What exactly is the impact of land use on transportation? This is

Table 2.4

National ITS Architecture Program Electronic Payment User Services Bundle

User Service	Notes
3.1 Electronic payment services	The use of payment systems technologies to enable travelers to pay without cash for tolls, transit tickets, and parking.

Table 2.5

National ITS Architecture Program Commercial Vehicle Operations User Services Bundle

User Service	Notes
4.1 Commercial vehicle electronic clearance	Services that support the automation of processes associated with commercial vehicle clearance.
4.2 Automated roadside safety inspection	Services that automate the roadside safety inspection process for trucks and loads.
4.3 On-board safety and security Monitoring	Services that rely on onboard equipment to monitor safety and security of trucks and loads.
4.4 Commercial vehicle administrative processes	Services that automate the commercial vehicle administrative process.
4.5 Hazardous materials security and incident response	Services that enhance the operation of hazardous materials security and incident response related to hazardous materials.
4.6 Freight mobility	Services that enhance the mobility of freight through maximizing trip time reliability and minimizing absolute trip time.

Table 2.6

National ITS Architecture Program Emergency Management User Services Bundle

User Service	Notes
5.1 Emergency notification and personal security	Services that enable travelers to notify the appropriate authorities in the event of a personal emergency our need for assistance.
5.2 Emergency vehicle management	Services that improve the efficiency of the emergency response process by minimizing response times and optimizing the use of available resources.
5.3 Disaster response and evacuation	Services improve the effectiveness of disaster response and evacuation with respect to the transportation system.

typically addressed through a series of surveys that lead to an estimate of the volume of trips generated by each land use. Land uses may be related to retail, educational, commercial, or manufacturing endeavors. We understand to a certain degree how transportation affects land use. For example, it has been observed in many cities across the world that the implementation of a metro system leads to accelerated development near the metro stations. Another ques-

Table 2.7
National ITS Architecture Program Advanced Vehicle Safety Systems User Services Bundle

User Service	Notes
6.1 Longitudinal collision avoidance	In-vehicle services that enhance cruise control with the ability to maintain a safe distance between the vehicle and the vehicle in front.
6.2 Lateral collision avoidance	In-vehicle services that improve safety by providing lane departure or lateral collision warning.
6.3 Intersection collision avoidance	In-vehicle services that can collaborate with roadside equipment to warn drivers of potential collisions at intersections.
6.4 Vision enhancement for crash avoidance	Services that provide enhanced vision for drivers in an effort to avoid crashes.
6.5 Safety readiness	Services that enable driver and vehicle condition to be monitored along with the condition of the roadway.
6.6 Precrash restraint deployment	Services that provide predeployment of safety devices and ancillary equipment that reduce the consequences of a crash.
6.7 Automated vehicle operation	Driverless vehicles that are capable of complete operation with no human intervention.

Table 2.8
National ITS Architecture Program Information Management User Services Bundle

User Service	Notes
7.1 Archived data	Services that support the use of data storage and management systems to extract data from intelligent transportation system applications and apply the data to planning and management of transportation.

Table 2.9
National ITS Architecture Program Maintenance and Construction Management User Services Bundle

User Service	Notes
8.1 Maintenance and construction operations	Services that enable maintenance and construction operations to take advantage of advanced technologies for fleet management, plant management, and overall site control

tion within this group would be, How good is our understanding of the effects of transportation on land use and exactly how does transportation influence land development?

What Are The Service Deficiencies in the Transportation System?

To identify and define service deficiencies it is necessary to establish the intended level of service for each mode. Typically, most city transportation providers do a good job at defining service level targets and measuring actual service levels achieved. Setting service level targets and measuring service levels achieved across the combined transportation system involving multiple modes offers scope for improvement.

What Is Current Demand and What Will Future Demand Be for Transportation?

This is really a question in two parts. In the first part the current demand for transportation in the city is addressed. This begs the question do we have sufficient data to characterize current demand? The obvious follow-up question is do we have sufficient capacity to meet current demand? It could be argued that congestion in any transportation mode is nothing more than an excess of demand over capacity. This argument assumes routine operating conditions, since congestion may also because by unexpected situations such as crashes, incidents, and vehicle breakdowns. The second part of the question relates to the future: Based on what we know of the past, is it possible for us to accurately predict demand into the future? The future can be defined as the next five minutes, the next hour, or the next 50 years. This leads to further questions concerning the tools that we have available to make predictions and how the accuracy of those tools is verified.

How Do We Maximize Access to Jobs?

In many smart city plans, a core objective is to maximize access to jobs. This leads to questions such as, Where do workers live? Where are suitable employment opportunities? An additional question would be, What current transportation options are there between jobs and work locations? This may also lead to questions such as, How can we use transportation to stimulate growth and job opportunities?

2.8   User Experience-Related Questions

User experience-related questions focus on customer service and how users perceive the quality of service in addition to actual measurements of performance.

How Do We Enhance the User Experience for Travelers in the Transportation System?

User experience has become an important parameter for business and commerce to drive business activities and guide investment decisions. It is highly likely that transportation will follow a similar path. In order to enhance the user experience, it is necessary to understand user behavior, identify what the user needs, and identify opportunities for providing new services that don't currently exist. As they say in the world of data science, the user is going to have an experience whether you like it or not, the question is will it be a good one and will it be one that you have engineered?

Are Transportation Customers Getting the Best Value for Their Money?

This question relates to a combination of travel speed, travel time, reliability, price, and levels of comfort. This could also be defined as a traveler stress index. Asking the question involves a comparison between different modes to determine price, reliability, and overall travel time. This may well reveal that certain sections of the community are not getting a fair deal.

How Do Transportation Customers Perceive Service Levels?

This question is typically answered through a range of surveys that take snapshots at specific times and locations. An alternative approach to asking this question, in the light of current data science capabilities, would be to ask this question across the entire city on a continuous basis. It is obvious that user sentiment changes with time and location and with events in the transportation system such as construction and service delays. Given that user expectations are being set by organizations not involved in transportation, such as Netflix and Walmart, it is also important to ask this question within the context of how well we are doing compared to other people. Not just other transportation people but how well are we doing compared to other people who are influencing customers. If your customer perception rating drops below that of an airline or an insurance company, then perhaps you should be worried.

What Impact Will Mobility as a Service Have on the Transportation System?

The impact of services such as Uber on society and traveler behavior is already clear. In many states the number of citations for driving under the influence have been significantly reduced because of the availability of the Uber service. As this is extended into a wider portfolio of options, then what impact will this have on our transportation system? This involves other questions such as how should transportation investments be targeted in the future and how can they be fully coordinated with private sector investments to achieve a portfolio of services.

How Can Travelers Make the Best Use of the Transportation System?

Good digital video recorders can be purchased at Walmart for less than $300. Purchasers who take them home would be absolutely appalled if they did not have a user manual in the box. In contrast, multibillion-dollar transportation systems are created in cities and very often don't have an adequate user manual. Asking this question leads to additional questions such as, What information do travelers need? What is the best way to deliver it to them? What is the best way to influence traveler behavior through the delivery of traveler information?

How Will Better Information Change Travel Behavior?

Posing this question requires a definition of "better" when it comes to traveler information. There are probably multiple dimensions to "better," including faster access to information, better quality information, and more complete and more timely information. If these can be delivered, then what changes can we expect in traveler behavior. How will decision quality information change traveler's choices when it comes to choice of route, choice of mode, and timing of the journey? These are important questions in the definition of strategies to manage capacity and demand for transportation.

2.9 What Do We Do with the Questions?

All of the questions that have been identified in the preceding text can be effectively and completely addressed using a data lake and analytics. It is hoped that defining and describing these questions will widen the scope of awareness of possible actions. This is important as questions are typically framed by a perception of what is possible. It would be valuable at this point to identify the questions that directly address your current needs, issues, problems, and objectives. It is also useful to start thinking about the opportunities and challenges that are related to answering the questions, so advice on how to address those is contained in Chapter 12.

References

[1] Palmisano, S. J., "A Smart Transportation System: Improving Mobility for the 21st Century, *Intelligent Transportation Society of America, 2010 Annual Meeting & Conference,* Houston, Texas, May 5, 2010. https://www.ibm.com/smarterplanet/global/files/us__en_us__transportation__ibm_samjpalmisano_smartertransportation_systems_05052010.pdf., retrieved July 27, 2016.

[2] "V2 V: Cars Communicating To Prevent Crashes, Deaths, Injuries," U.S. Department of Transportation website, https://www.transportation.gov/fastlane/v2v-cars-communicating-prevent-crashes-deaths-injuries retrieved July 27, 2016.

[3] The National Intelligent Transportation Systems Development Program, http://www.iteris.com/itsarch/html/user/userserv.htm. retrieved on July 27, 2016.

3

What Is Big Data?

3.1 Informational Objectives of This Chapter

This chapter is designed to answer the following questions:

- How is big data measured?
- What are the attributes of big data?
- What has changed in the world of data science?
- What are some examples of big data from industry and commerce?
- What are some examples of big data from transportation?

3.2 Chapter Word Cloud

Figure 3.1 presents a word cloud for Chapter 3.

3.3 Introduction

Big data may potentially be as important to business and transportation as the Internet has been. More data leads to more accurate analyses and greater understanding of the underlying mechanisms that affect the operation of enterprises. This chapter explores the nature of big data based on best practices from data

Figure 3.1   Chapter 3 word cloud.

science. It then brings the subject to life from a transportation perspective by providing some examples of big data sources for transportation. Using the right approach, the true value of big data can be realized. A good way to explain it is that data itself is difficult to handle and not very attractive and that its meaning is well hidden. It's a lot like the data that forms the basis for music. The notes in the page and the various musical frequencies come together to form something that is pleasant to the year as a result of arrangement and orchestration. Data is brought to life when additional tools and expertise are applied to extract the desired result. Figure 3.2 illustrates how data and musical notation is converted to sound by a musician playing a musical instrument resulting in palatable information in the form of music.

3.4   How Is Big Data Measured?

In the discussion regarding the nature of big data, beginning in Section 3.5, big data size terms are used to indicate the relative size of big data in various transportation and nontransportation applications. Therefore, before starting to explore big data, it is useful to include a discussion of how big data is measured.

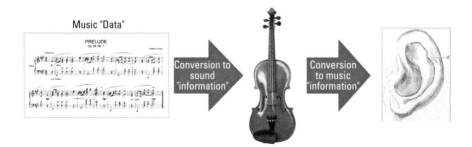

Figure 3.2   Data to information—a musical analogy.

Like all data, big data is measured in bytes. A byte is eight bits, and a bit represents either a one or a zero, otherwise known as binary or numbers to the base 2. This numbering system is useful because switches in a computer can be turned off to represent 0 and on to represent one. The term bit is short for binary digit and can be considered the smallest unit of data in a computer. You might also like to know that half a byte is called a nibble. In telecommunications and Internet traffic, the speed of the connection is usually described as the number of bits per second. This is often referred to as *bandwidth,* although it describes the speed rather than the capacity of the communication channel. An ASCII character, one of the characters used to build the words in this sentence, consists of one byte.

As data sizes have grown, prefixes have been placed in front of the byte to represent multiples of 1,000. Table 3.1 [1] shows what these prefixes mean in terms of the number of bytes; and to provide some context, it includes notes to relate the sizes to something recognizable. For example, if you know that a data size is 4 Zb, it is hard to comprehend. If you know that annual global Internet traffic is expected to pass the zettabyte threshold by the end of 2016 and will reach 2.3 Zb by 2020 [2] then you can start to comprehend the enormity of the number.

3.5 What Is Big Data?

Big data is a widely used term, and it is important that we have a clear and commonly agreed on definition of the subject. Big data describes data sets so large and complex that they become difficult to process using conventional data-processing hardware and software. There is a trend toward assembling larger

Table 3.1
Orders of Magnitude of Data [1]

Size and prefix	Base 10	Notes	References
1,000 bytes = 1 kilobyte	3		
1,000 kilobytes = 1 megabyte	6	A typical English book volume in plain text	[3]
1,000 megabytes = 1 gigabyte	9		
1,000 gigabytes = 1 terabyte	12	212 DVDs	[1]
1,000 terabytes = 1 petabyte	15		
1,000 petabytes = 1 exabyte	18		
1,000 exabytes = 1 zettabyte	21	Internet traffic in 2016	[2]
1,000 zettabytes = 1 yottabyte	24		
1,000 yottabytes = 1 brontobyte	27		
1,000 brontobytes = 1 geopbyte	30		

data sets because these are richer sources of insights and understanding. A large data set allows an enterprise-wide or organization-wide view that can yield more information than a series of silos or smaller data sets.

Big data can be considered to be an evolution of data science with some aspects that are new and some are not. For example, most potential transportation big data applications address safety, efficiency, and enhanced user experience. These are issues that the transportation profession has been addressing for a number of years. Aspects that are new include exponential growth in data sizes and new availability of data—both structured and unstructured. This combines with rapid acceleration in many dimensions (volume, velocity, variety, variability, and complexity).

Other new aspects featured by big data include the following:

- *Analytics:* The ability to conduct graph and path analytics, and analytics on new, nonrelational data types coupled with existing relational data.

- *Tools:* New tools that can help to uncover insights from data such as text in accident reports or patterns in visuals, to quickly find the signal in the noise.

- *Economics:* New capabilities with reduced cost mean that data can be retained. It is not necessary to throw away signal timings, speed, flow, and occupancy data. By leveraging new techniques, it is possible to apply the appropriate storage mechanism in terms of cost and performance to the appropriate data set. This also enables appropriate access to the different data types.

- *Architecture:* The emergence of a hybrid ecosystem that allows both old and new tools to work together within a single framework to enable rapid discovery analytics on new data.

My first exposure to the term big data in 2011 sparked an interest in how long the term had been in use. Subsequent research on the origin of the term uncovered that it can be attributed to one of two people (according to the *New York Times* [4]). Anecdotal evidence suggests that it was first introduced by John Massey from Silicon Graphics in the mid 1990s. He wanted to use a single term to describe a range of issues in data storage and data management. The other possible author of the term is John Diebold of the University of Pennsylvania, who first used the term in association with macroeconomics in his paper *Big Data Dynamic Factor Models for Macroeconomic Measurement and Forecasting,* which was first presented in 2000 and published in 2003. Today the term has to come to represent not just volume of data but also a range of dimensions, listed as follows:

- Type;
- Volume;
- Velocity;
- Variety;
- Variability;
- Complexity;
- Veracity.

Let's take a look at each of these in turn.

Type

There are two major categories of data: real-time and archive. The literature indicates that these are given many different names; for example, real-time data may be referred to as transactional and archive data may be referred to as static data. They are often referred to as "hot" and "cold" data, giving the sense that hot data is live and used in the short term while cold data is stored for longer-term use. The terms *data at rest* and *data in motion* are also used to differentiate static and dynamic data. The distinction lies in how the data is being used at any given time. Real-time data must be kept in a manner that is accessible quickly. To support this, less frequently used data can be moved to an archive where the data can be stored in large volumes for long periods of time at lower cost. These days it is also possible to conduct analytics on a real-time data stream while it's on the way to being stored. The use of real-time analytics is another reason for separating real-time data from archive data.

Volume

The volume dimension of big data is an obvious one. The adjective *big* gives you the sense that this part of data science is about volume. In the past, there has been a tendency to fragment bigger data sets to store data more efficiently and enable fast access. These days, with the advent of fast and low-cost data storage, the tendency is to consolidate and bring data to a central repository. This has the effect of creating an enterprise-wide view of the data, which could be difficult if the data is fragmented and stored in silos across the organization. So how big is big data? Here are a few examples, from beyond transportation:

- Approximately 1 Pb of data is uploaded to YouTube every day [5].
- It is estimated that the human brain has a functional memory capacity of 2.5 Pb [6].
- Netflix users stream approximately 4.7 Pb of data every year [7, 8].

I'll contrast these numbers to an experience at the London Borough of Hackney in the early 1980s when a Superbrain personal computer with a 2-Mb hard disk was acquired. Staff and management were not sure what they were going to do with all that storage space! Note that a megabyte is a billionth of a petabyte.

Velocity

It would be simple to equate the velocity at which data arrives as the number of bits per second that a telecommunications link can support. For example, a typical Internet connection for business might support data speeds of around 100 Mbps. However, the velocity dimension goes beyond the speed at which communications can be achieved to address a wider measurement of the speed at which we can go from sensing data to doing something about it. While data communications throughput and latency can measure the speed at which data can flow across the network, there are other factors that will influence the speed at which we can make use of the data and turn it into information. As shown in Figure 3.3, there are several steps in this process, addressed as follows.

- *Sense:* In this step the data is collected by several different means including sensors, closed-circuit TV cameras, smart phones, and roadside infrastructure-based sensors. The data can also include anecdotal data and can be structured or unstructured.

- *Ingest:* This is the process step in which the data is assembled into a data platform and brought together into a meaningful and coherent body of data. This topic is discussed in Chapter 9.

- *Process:* In this step the raw data is turned into information. The huge volume of zeros and ones and raw signal data is converted into meaningful summaries, tables, visualizations, and other such structures that allow humans to understand the trends and patterns within the data.

- *Assimilate:* In this step of the process, the humans involved assimilate the information and begin to think about the impact that the information will have on their job and their organizations. This may well lead to the recognition that further information is required. It may also highlight the need for organizational fine-tuning to ensure that the assimilation process is as efficient as it can be.

Sense ⮕ Ingest ⮕ Process ⮕ Assimilate ⮕ Understand ⮕ Act

Figure 3.3 Steps from sensing to action.

- *Understand:* The development of knowledge based on information received such as an understanding of the underlying mechanisms that drive transportation demand or a detailed picture of traveler behavior. This may lead to what could be referred to as *wow* and *whoops* moments. A *wow* moment occurs when a connection or mechanism that was not previously obvious is identified and understood. A *whoops* moment is when new understanding brings a realization that there are deficiencies in the delivery of current transportation services. A *whoops* moment does not necessarily represent a catastrophic event. A good response to a *whoops* moment would be to recognize that a new problem has been identified and develop a plan with the associated budget to fix it. This is this step into the realm of scientific investment planning or results-based work program development, which is discussed in more detail in Chapter 8.

- *Act:* The final step in a process involves taking the understanding that has been gained using the data to create new information and turn it into actionable strategies. It is important to consider this step even at the beginning of the process when data is being collected. There would be no point in collecting data and spending time converting it into information, if it resulted in understanding that could not be put into action. At our current stage in smart city transportation this could involve the reassignment of resources, the definition of new capacity requirements, the application of more effective traffic engineering, or a change in transit service frequency.

It is noticeable that many transportation agencies embark on large-scale data collection exercises without a clear and detailed understanding of the use to which the data will be put. It is hoped that the application of big data and analytics techniques will encourage transportation agencies to take a wider view of the process and develop a detailed understanding of the proposed uses of the data. This will in turn lead to better approaches for defining the quality of the data required. In the longer term, it could also be expected that the establishment and operation of data lakes (centralized repositories of data) will also enable automated responses as we develop a better understanding of cause and effect in transportation. Chapter 9 details data lakes. Figure 3.4 illustrates a trend that is anticipated in industry and commerce as big data and analytics pervade the organization. It is to be expected that a similar trend in transportation will emerge.

From a starting point of reporting, better capabilities to analyze the mechanisms that relate to transportation demand and supply enable us to build the capability to predict future transportation demand, supply, and conditions.

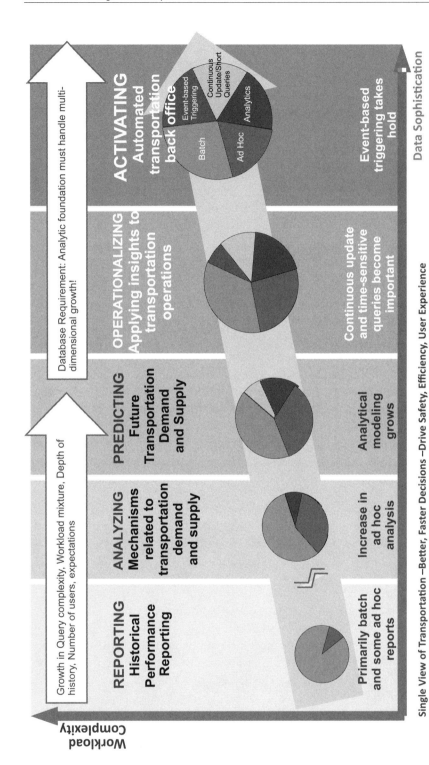

Figure 3.4   The pathway from reporting to automation.

This leads to the ability to operationalize the knowledge by applying the insights that have been gained regarding transportation. This in turn leads to the ability to automatically activate events that are based on triggers that have been defined as a result of the analytics. This incremental growth is enabled by a parallel migration in which ad hoc analysis increases and analytical modeling grows, followed by the ability to continuously update and answer time sensitive queries. Ultimately event-based triggering becomes a reality. In transportation terms event-based triggering could mean automatically generating messages to be displayed on dynamic message signs or the automatic retiming of traffic signals based on fluctuations in demand and the occurrence of events. It could also include automated response to incidents including the dispatching of resources and the use of traffic control devices to manage traffic.

Variety

Another dimension of big data is an increase in the variety of data that can be collected and processed. For example, in the world of transportation, it is possible to collect a wide range of data including the following:

- Traffic speed;
- Traffic volume;
- Travel times;
- Transit passenger counts;
- Vehicle location.

The above items can be described as traditional transportation data. In the era of big data, when we have no need to constrain data sizes, we can also add the following nontraditional data:

- Electricity consumption;
- Retail transactions;
- Smart phone position data;
- Probe data from connected vehicles.

These are just a few examples of additional data that can be blended with traditional data to create big data with a wider variety of data. In fact, the surface has just been scratched in terms of the wide variety of data that can be collected to gain better insights into the demand for transportation including the volumes of and the reasons for travel. It is to be expected that another dimension of variability will lie in the expansion of data collection within transportation agencies. It is typical that the central focus for data collection in a transportation agency lies in the operations department. As Figure 3.5 illustrates, there is much more to transportation than operations.

Note that different transportation agencies may have different terminology used to describe each of these stages and that Figure 3.5 uses generic terms. In fact, one of the challenges facing transportation agencies in the light of big data and analytics will be how to share data and information across the entire organization in a seamless manner.

Variability

Data analysts and data scientists have an interesting perspective on variability. They love it. A typical approach to variability is to create averages and summaries to be able to handle it. The data experts see this as a problem because their perspective is that the value is in the detail and that summaries and averages remove detail. A good example in transportation would be the average travel time between traffic signals. While this is a useful measure of performance, it ignores detail that would be extremely valuable. For example, if a driver travels at 50 mph and then stops at a red light, then travels at onward at 50 mph and then stops at the next red light, it could be measured that on average the vehicle traveled at 30 mph. This journey is not differentiated from another driver who might experience a smooth, steady 30-mph journey through the corridor, receiving green signals at each intersection. Averaging removes some important detail. The use of detailed, second-by-second speed profiles for individual vehicles traveling along a signalized corridor would address this issue. Such profiles have, in the past, been considered to be unmanageable. It is only recently that data collection techniques such as probe vehicle data collection have enabled the acquisition of such data. In the world of big data and analytics the data is available, *and* the horsepower to convert it to meaningful information is at hand.

Complexity

Data is becoming more complex, and the ability to capture the same data from more than one data source adds to this complexity. Techniques have been developed to compare the same data from multiple data sources—this is known as octagonal sensing. Figure 3.6 provides an information technology–centric view of all the factors that go along with increasing data complexity.

Figure 3.6 shows a transition from enterprise resource planning (ERP) that addresses the optimization of assets within an organization to customer relationship management (CRM) that considers the external interface between the organization and the customer. This leads on to extensive use of the web, which is not only driven by big data but can itself generate big data. This migration from ERP to big data features increasing data variety and complexity but

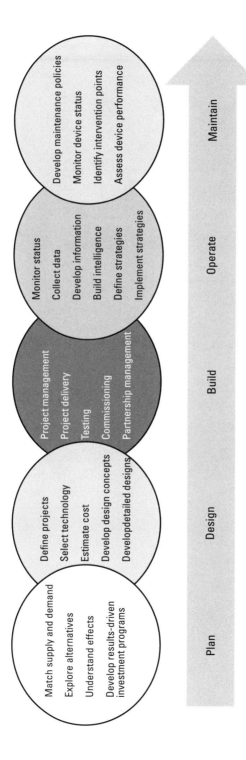

Figure 3.5 The full spectrum of transportation delivery activities.

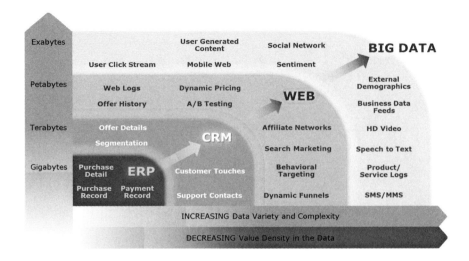

Figure 3.6   The growth of information complexity.

decreasing value density in the data. Value density is a way to measure the value that you would obtain from data relative to the size of the data. The reduction in value density is another reason for creating a data lake in order to be able to extract maximum value from the data.

Veracity

This dimension relates to the uncertainty associated with big data. As the speed at which data arrives increases and the variety grows to include structured and unstructured data, there can be a significant amount of uncertainty relating to the precision of the data. This requires tools and techniques that enable us to understand the uncertainty in the data and, perhaps, to apply rankings to the quality of the data. The veracity or ability to trust the data might also be impacted by tampering or other issues that would affect the quality of the data. This adds a security dimension to veracity.

3.6 Challenges

In addition to the opportunities, there are challenges associated with big data, listed as follows:

- Complexity analysis;
- Capture;
- Curation;

- Search;
- Sharing;
- Storage;
- Transfer.

In describing these challenges from a transportation perspective, it is possible to offend the data analysts and data scientists who will read this book, since a simplistic view has been adopted. However, no offense is intended. The focus is explaining the value to transportation rather than developing a technical description of the subject. The objective is to provide an awareness of the challenges, to illustrate their nature and, to provide an overview of how they are addressed in data science.

Complexity Analysis

This is an emerging field in data analysis and data science that categorizes data according to its complexity. As data sets rapidly increase in scale, and processing becomes automatic, multiple systems can be connected together; this leads to increasing complexity. If this is left unmanaged, it can lead to unpredictable behavior within the system and difficulties in processing the data. A typical engineering approach would attempt to remove the complexity, but this runs counter to obtaining maximum value from big data. As discussed earlier in Section 3.5, the real value lies in the detail, so complexity cannot be avoided. Tools and techniques have been developed in the field of complexity analysis that enable the understanding of complexity and the development of new approaches to modeling and controlling complexity in systems.

Capture

Relative to big data, data capture represents another challenge. While the transportation community is adept at capturing automated data from sensors and other roadside devices, the world of big data requires that multiple data sets be combined to give us the insights that we're looking for. This means that unless the amount of resources we invest in data capture is expanded, automated solutions must be considered. Data capture includes the process of bringing the data back to a central repository and the work required to bring the data into the repository. In the data world this is referred to as extraction, transformation, and loading (ETL). If the multiple data sources include data from beyond the organization, then the data capture process will also include the establishment of some form of data-sharing agreement.

Curation

One of the major challenges facing transportation with respect to current data, never mind Big Data, is to have an organization-wide awareness of what data is available and where it is located. Today, there are automated tools available to assist in the curation process and provide support for master data management and the definition of data governance techniques. One of the important elements of data creation lies in the definition and specification of who can have access to the data and what permissions they have to alter the data.

In order to emphasize the importance of addressing this challenge effectively, here are some anecdotes from the field. The Florida Department of Transportation provided the following real-life stories in a recent presentation [9]:

- A study on the feasibility of a new road was completed by a project team. Six months later another project team performed the same study. They did not know of the existence of the first study and the associated data.

- The project team was delayed and moving on to the next phase of a road project as they were unable to locate survey data. Due to schedule requirements the team had to repeat the survey and recollect the data.

- A project to resurface a particular stretch of road was completed. One month later, a new street lighting system was installed on the same stretch of road, requiring that the road surface be excavated.

These stories capture the essence of data curation and the challenges it poses from a transportation point of view. There is little point in investing resources for data collection if the organization is not aware of the data's existence and cannot access the data in a timely manner. These insights led the Florida Department of Transportation to embark on a major program to equip the organization with the latest data management technology.

Search

To be useful to transportation professionals, search tools and techniques must be intuitive and relatively simple to understand. It is important that the emphasis is placed on the use and value of the tools rather than the need to develop skills and expertise to make use of them. Users have grown accustomed to the power of Google, and this provides excellent access to search the World Wide Web for data. It is essential that the tools used to search big data also feature these intuitive capabilities. The sophistication of search tools is growing, enabling the presentation of more precise results rather than simply a list of possibilities. This is not only relevant to the World Wide Web but should also be applied to internal big data systems.

Sharing

Data sharing has the potential to be one of the biggest challenges facing transportation agencies. The physical ability to share data over telecommunications networks utilizing fiber optics and copper wire is just the beginning. It has been learned that unless there is also a data-sharing agreement in place then nothing will be transmitted across the telecommunications network. Data sharing can also be a challenge between the public sector and the private sector. The definition of suitable sharing agreements is paramount to enabling data flow. To effectively address data sharing issues, it is also necessary to develop a business model for the data. One business model could be that the public sector simply collects the data as part of its everyday operations to deliver effective transportation and then provides this data free to private sector organizations that wish to make it the basis for their products and services. Under another approach, public sector agencies would attempt to develop agreements to enable them to share in the value of the products and services that are derived from public sector data. The definition of a business model for data would also include the detailed definition of how much data processing is conducted within a public agency and the cost of data collection and processing.

Storage

The data storage challenge revolves around the management of data sets across data storage infrastructure. Unless these infrastructures are properly managed and structured, it is possible to spend an undue amount of time and money on data storage. This takes us back to the distinction between real-time data and archive data. One of the obvious ways to address the structure of data storage would be to separate them so that the more expensive real-time ways to store and access data are used on the appropriate data set. It is interesting to note that this problem is being driven by an extremely positive development in the market—the cost of acquiring data storage has been dropping dramatically. However, this has led to hidden costs beyond acquisition when it comes to maintaining and managing the data. Figure 3.7 illustrates the reduction in the cost to acquire data storage, showing the average hard drive cost per gigabyte from 1980 to 2009 [10].

Over time, the cost of hard drive storage acquisition has dropped significantly from a peak value $700,000 per gigabyte to a low of $0.3 per gigabyte. Of course, acquiring the hard disk space is only one element in the cost of data storage. Other elements include maintenance, upgrades, power, physical facilities for hosting the hard drive storage, and the cost of managing it. However, the cost of hard drive storage acquisition seems to be a good yardstick to indicate how dramatic the cost reduction has been. Another option that has emerged in recent times involves simply renting or leasing data storage space at

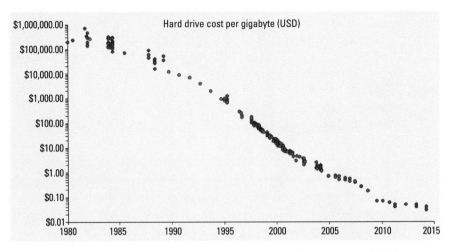

Figure 3.7 The dramatic reduction in hard drive storage cost from 1990 to 2009 [10].

someone else's data center or in the so-called cloud. The cloud is a network of servers with some servers processing data and some servers storing data.

Transfer

When a big data set gets into the high terabytes or low petabytes, data transfer becomes an issue. The following Table 3.2 shows data transfer times for a petabyte and a petabyte of data at the various transmission speeds that might be available to a transportation agency.

As Table 3.2 shows, even at the fastest speed of 10,000 Mbps (10 Gbps) it takes almost 14 minutes to transmit a petabyte of data. This suggests that co-location of various data ingestion and processing hubs should be given serious consideration in the development of a big data and analytics capability.

3.7 Big Data in Transportation

To conclude this chapter, we bring big data to life from a transportation perspective by exploring some potential sources of big data within the world of transportation. Note that the definition of big data as discussed earlier is utilized in this chapter and that it's not necessarily all about volume, but also variety and velocity. The intent is not to create a catalog of all possible sources of big data in transportation, but to provide an overview of the possibilities. Big Data in transportation can be viewed from the perspective of the different application areas for advanced technologies in transportation. These are listed in Table 3.3, then discussed one by one.

Table 3.2
Data Transmission Times

Data Transmission Rate (Mbps)	Data Transmission Times in Minutes	
	Petabyte	Terabyte
10	13,333.33	13.3
50	2,666.67	2.7
100	1,333.33	1.3
1,000	133.33	0.1
10,000	13.33	0.0

Table 3.3
Advanced Transportation Technology Application Areas

Traffic management
Traveler information
Public transportation management
Electronic payment
Commercial vehicle operations
Emergency management
Connected and autonomous vehicles
Smart cities
Archive data
Maintenance and construction operations
Performance management

Traffic Management

Included in this category are all activities within the traffic management area including traffic signal control, freeway, and incident management. Table 3.4 lists data that could be generated from these sources, representing a sample of data possibilities within traffic management. A more complete catalog is available in the form of the *Traffic Management Data Dictionary* [11].

Traveler Information

Traveler information involves collecting data about the transportation system and converting it into information that can be delivered in a timely manner to travelers. The overall concept is to improve travel decision-making through the delivery of decision quality information. Traveler information can be delivered through multiple channels including the web and in vehicle information systems. Table 3.5 summarizes data that could be sourced from a traveler information system.

Table 3.4
Traffic Management Data Sources

Traffic Signal Timing
Traffic speed
Traffic volume
Intersection turning movement counts
Travel times between intersections
Number of incidents
Duration of incidents
Response times for incidents

Table 3.5
Traveler Information Data Sources

Traveler Information
Demand for travel
Travel conditions
User preferences
Cost of travel
Reliability of travel time
Availability of travel options
Best route options
Ride matching options

Public Transportation Management

Public transportation management includes activities required to manage the fleet of transit vehicles and provide passenger information. It includes the management of fixed route, paratransit, and flexible route services. Fixed-route services follow a predefined route and schedule. Paratransit services are typically on-demand and available to transportation disadvantaged customers who cannot travel by the other public transportation services. In recent times flexible route services have been introduced that combine some aspects of fixed-route, with some aspects of on-demand. For example, one approach to flexible-route services would be to assign a bus within the zone or region and allow customers to call the bus service on-demand. The customer would be able to request bus service from the residence to any point within the zone or to the closest access point to fixed-route services. Table 3.6 lists data possibilities.

Electronic Payment

These are designed to make it easier for travelers to pay for transportation services. They include electronic toll collection, electronic ticketing for transit, and

Table 3.6
Public Transportation Management Data Sources

Public Transportation Management
Vehicle location
Current service levels
Passenger loads
Number of passengers boarding and alighting at various bus stop or station
Demand for transit services
Availability of travel options
Best route options
Ride matching options
Staff resource allocation

electronic payment for parking fees. The systems can also be significant generators of valuable data, including the data items shown in Table 3.7.

Commercial Vehicle Operations

Commercial vehicle operations address the management and operation of truck-based freight systems. This involves the management of the truck and the driver and the administrative processes associated with operating a truck freight business. Table 3.8 lists data items that could be generated by commercial vehicle operations systems.

Emergency Management

Emergency management addresses the fleet management aspects of emergency response vehicles, the delivery of emergency notification alerts, and the activi-

Table 3.7
Electronic Payment Data Sources

Electronic Payment
Passenger loads
Demand for transit
Number of passengers boarding and alighting at various bus stop for stations
Fares
Revenue
High-occupancy vehicle and express lane use
Transactions
Accounts

Table 3.8
Commercial Vehicle Operations Data Sources

Commercial Vehicle Operations
Driver credentials
Truck credentials
Load credentials
Vehicle condition
Driver condition
Load condition

ties associated with disaster response and evacuation. Table 3.9 lists a sample of data that could be generated by an emergency management system.

Connected and Autonomous Vehicles

The connected vehicle involves a supportive a two-way dialogue between vehicles and back-office infrastructure and from vehicle to vehicle. Chapter 4 delves further into this topic in its discussion of connected and autonomous vehicles. The ability to extract data from vehicle systems enables the sample data possibilities listed in Table 3.10.

Smart Cities

The transportation elements of a smart city offer considerable data generation possibilities that include all of the other application areas described in this chapter. A smart city will have a focus on connected citizens and connected visitors. Table 3.11 lists additional data possibilities that could be enabled by a smart city.

Table 3.9
Emergency Management Data Sources

Emergency Management
Emergency vehicle location
Location and nature of emergencies
Threat data
Transportation network status
Disaster response plans
Incident management plans
Hazardous materials incident response plans

Table 3.10
Connected and Autonomous Vehicles Data Sources

Connected and Autonomous Vehicles
Vehicle operating data
Vehicle location
Instantaneous vehicle speed
Vehicle ID
Driver status
Driving behavior

Table 3.11
Smart Cities Additional Data Sources

Smart Cities
Movement analytics based on smart phone position data
Crowd-sourced data service
Social sentiment and user perception

Archive Data

In recognition of the data possibilities associated with the application of advanced technologies to transportation, the U.S. Department of Transportation in its National ITS Architecture Development Program introduced this user service category. It defines the extraction of data from all the other application areas for the purposes of system and performance management.

Table 3.12 lists a sample of data possibilities.

Maintenance and Construction Operations

Maintenance and construction operations involve the activities associated with updating and creating transportation infrastructure. This includes asset management and the various aspects of project management involved in construction and implementation. Table 3.13 lists a sample of the data possibilities.

3.8   Transportation Systems Management and Operations

Data possibilities have been grouped into application areas because data tends to be siloed into these categories at the current time. Transportation consists of a series of specialized focus areas within which specific data is collected and utilized. It has also been observed that in many cases the data coming into a transportation organization emanates from operations and may not necessarily find its way to other departments, such as planning and design. This is

Table 3.12
Archive Data Sources

Archive Data
Current operating conditions
Device status
Asset
Traffic
Transit
Traveler information
Public transportation management
Electronic payment
Commercial vehicle operation
Emergency management
Connected and autonomous vehicles
Maintenance and construction operations

Table 3.13
Maintenance and Construction Operations Data Sources

Maintenance and Construction Operations
Stages of construction activities
Alternate route and detours
Work zone speed limits
Asset condition
Asset location and inventory
Construction resource assignment
Construction work zone status
Labor resource allocation

something that is expected to change dramatically as the concept of big data is introduced. Indeed, the transportation profession is already moving in this direction with the introduction of the transportation systems management and operations (TSM&O) concept, which is intended to span the various activities. For example, travel time reliability and service levels in public transportation can be significantly impacted by the quality of traffic management. The progress made by a fixed-route transit bus is directly related to traffic congestion and other traffic conditions. TSM&O offers the possibility of closer integration between planning, design, building, operating, and maintaining transportation systems. The implementation of big data will be an extremely valuable tool in this integration and coordination.

References

[1] Wikipedia, https://en.wikipedia.org/wiki/Terabyte#Illustrative_usage_examples, retrieved April 3, 2017.

[2] Cisco website, http://www.cisco.com/c/en/us/solutions/collateral/service-provider/visual-networking-index-vni/vni-hyperconnectivity-wp.html, retrieved July 30, 2016.

[3] Wikipedia, https://en.wikipedia.org/wiki/Megabyte#Examples_of_use, retrieved April 3, 2017.

[4] *New York Times* "Bits" blog, http://bits.blogs.nytimes.com/2013/02/01/the-origins-of-big-data-an-etymological-detective-story/, retrieved July 29, 2016.

[5] Google Cloud Platform blog, https://cloudplatform.googleblog.com/2016/02/Google-seeks-new-disks-for-data-centers.html, retrieved July 29, 2016.

[6] Scientific American.com, "What Is the Memory Capacity of the Human Brain?" Paul Reber (Professor of Psychology at Northwestern University), https://www.scientificamerican.com/article/what-is-the-memory-capacity/, retrieved April 3.

[7] Netflix help center, https://help.netflix.com/en/node/87, retrieved on July 29, 2016 at 3:07 p.m.

[8] Netflix quarter to 2016 letter to shareholders, http://files.shareholder.com/downloads/NFLX/2531277189x0x900152/4D4F0167-4BE2-4DC1-ACC7-759F1561CD59/Q216LettertoShareholders_FINAL_w_Tables.pdf, retrieved on July 30, 2016.

[9] Information Technology Strategic Plan Update, Enterprise Information Management, ROADS project, presentation delivered by April Blackburn, Chief Information Officer, Florida Department of Transportation, June 17, 2015.

[10] "History of Storage Costs," [online] updated March 9, 2014, by Matthew Komorowski, retrieved on August 1, 2016.

[11] *The Traffic Management Data Dictionary,* the Institute of Transportation Engineers, http://www.ite.org/standards/tmdd/, retrieved April 3, 2017.

4

Connected and Autonomous Vehicles

4.1   Informational Objectives

This chapter aims to answer the following questions:

- How is electronics and information technology affecting the automobile?
- What is a connected vehicle?
- What are the challenges associated with connected vehicles?
- What is an autonomous vehicle?
- What are the challenges associated with autonomous vehicles?
- What are the differences between connected and autonomous vehicles?
- How do connected and autonomous vehicles fit within a smart city?
- How might connected and autonomous vehicles affect transportation?
- What are the big data and analytics aspects of connected and autonomous vehicles?

4.2   Chapter Word Cloud

The word cloud shown in Figure 4.1 provides an overview of the content of the chapter by listing the most frequently used words, with the font size proportional to the frequency of use of the word.

Figure 4.1 Chapter 4 word cloud.

4.3   Introduction

Connected and autonomous vehicles fit within the bigger picture of the application of advanced technologies to vehicles and the treatment of the vehicle and treatment of the road as a single system. As depicted in Figure 4.2, the electronics content in private cars has been growing steadily since the 1950s. [1].

Today, between 30 and 35% of the total cost of a car is comprised of automotive electronics, and this figure is expected to grow to 50% by 2030. The electronics content of the vehicle represents a range of technologies summarized in Table 4.1 [2].

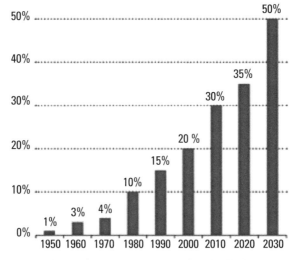

Figure 4.2 Automotive electronics as a percentage of total vehicle cost.

Table 4.1
Automotive Electronics

Automotive Electronics	Description
Active safety	Avoid and mitigate the effects of a crash
Chassis electronics	Monitor and manage the chassis
Driver assistance	Decision support for the driver
Engine electronics	Monitor and manage engine operation
Entertainment systems	In-car entertainment systems such as radio and digital music players
Passenger comfort	The air conditioning, heated seats, and other applications to increase passenger comfort
Transmission electronics	Monitor and manage the operation of the transmission between the engine and the wheels

Driver assistance and active safety systems can be considered to be part of the connected vehicle. Connected vehicles can be considered an evolution in the electronics content of the vehicle. They can also be conceived as part of a shift in the way vehicles and roads are perceived. For some time now there has been a growing realization that perhaps it would be better if we treated transportation as a single system.

It would appear that we are approaching a tipping point in transportation service delivery that will see a surge in connectivity and enable us to develop a much better understanding of current operating conditions and forecasted operating conditions for the near future. This in turn will generate significant volumes of data and the tremendous need to convert this data to information and understand the new insights that will be available. Big data and analytics will play a vital role in this.

The connected vehicle can be viewed as one element in the ultimate connected transportation system, through which the vehicle is connected to a back-office infrastructure and to other vehicles.

This chapter is not intended to provide a detailed technical exposition of the in-vehicle and telecommunication technologies that support connected and autonomous vehicles. Its objective is to provide an overview that explains the essential characteristics of connected and autonomous vehicles within the context of smart cities, big data, and analytics. An understanding of the essentials of connected and autonomous vehicles and the service evolution related to them is of significance when considering their potential for big data and analytics applications and the wider impact that they are likely to have on transportation service delivery.

The subject of this chapter is both *connected* and *autonomous vehicles*. At first glance, the terms can be quite confusing. While both types of vehicle fall under a single banner known as telematics, there are major differences between the two vehicle types. Nevertheless, both types are likely to have a significant impact on our transportation system and play a central role within a smart city, while providing new data in the form of probe vehicle data, hence the interest in explaining both concepts under the auspices of this book.

It is difficult to address big data and analytics and transportation without taking full account of both connected and autonomous vehicles. Connected and autonomous vehicles are a major focus for both federal and private sector investment. With the connected vehicle, the ability to establish a reliable two-way communication channel between the driver and an information technology infrastructure holds the promise of substantial advances in safety, efficiency, and user experience. There is more than one way for technology to support the two-way communication link with different approaches having different capabilities and suitability for different applications.

With respect to the autonomous vehicle there are potential gains in safety and user experience by introducing vehicles that are capable of driving themselves. This could be particularly useful regarding freight and transit, where the driver is not attempting to reach a particular destination but is present as a requirement for vehicle operation. The autonomous vehicle—discussed in more detail in Section 4.6—is likely to be introduced on a phased basis with emigration from driver support and assistance to full autonomous operation.

4.4   What Is a Connected Vehicle?

The connected vehicle essentially involves the use of wireless telecommunications along with in-vehicle equipment to support a two-way data exchange between the vehicle and the back office. There are two major approaches to the establishment and operation of the wireless link between the vehicle and back office. These can be characterized as follows:

- Wide-area wireless;
- Dedicated short-range communications (DSRC).

The wide-area wireless approach to the connected vehicle, involves the use of wide-area wireless via cellular wireless services to enable the two-way communication link between the back office and the vehicle. In this case the infrastructure to support communications to and from the vehicle is already in place in the form of wireless networks developed and operated by major carriers such as Verizon, AT&T, Sprint, and T-Mobile. These wireless services are utilized to support cloud-based back office applications. Figure 4.3 illustrates this cellular wireless approach to the connected vehicle with private-sector data and

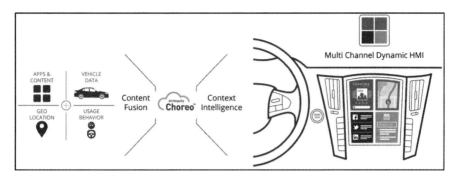

Figure 4.3 Overview of the cloud-based connected vehicle.

information making use of broadband wireless connections between the vehicle and cloud-based services.

Figure 4.3 shows the Airbiquity Choreo application [4] that provides a link between connected vehicles and automotive manufacturers through a cloud-based service. With this application, vehicle data, usage behavior, and geolocation along with apps and content are combined or fused to provide a data stream from the vehicle. This can also support data delivery that can be converted to information using a multichannel dynamic human machine interface within the vehicle (otherwise known as an in-vehicle information terminal).

The DSRC approach, as its name suggests, makes use of a dedicated communication link between the connected vehicle and roadside infrastructure. This approach builds on years of experience that have been accumulated in the application of electronic toll collection systems. The advantage of this approach is that the technology can support high-speed, high-reliability, and low-latency communications to and from the vehicle. The disadvantage of this approach is the need to install roadside equipment (RSE) at regular intervals along roads. The short-range communications provided typically enable vehicles to talk to an RSE unit over a range of about a kilometer. The approach also requires the installation of special equipment in the vehicle to enable the communications with the roadside infrastructure.

A narrow range of wireless telecommunications frequencies (spectrum band) was allocated for DSRC by the Federal Communications Commission (FCC) in October 1999. The band spans defined 75 MHz from 5.8502 to 5.925 GHz and lies in the microwave portion of the radio spectrum. While this band was initially reserved to promote the safety of life of automobile drivers, passengers, and pedestrians using DSRC the FCC is now considering how this spectrum can be shared by other users. Radio spectrum is a scarce resource, and growth in the use of Wi-Fi has caused a surge in demand for spectrum, causing the FCC to reevaluate spectrum allocation. However, sharing the band with other unlicensed applications could place a question mark on the reliability of DSRC.

Figure 4.4 illustrates the application of DSRC at an intersection showing vehicle-to-vehicle and vehicle-to-infrastructure communications to improve safety [5].

4.5 Connected Vehicle Challenges

There are a number of challenges associated with the connected vehicle. These challenges are described in the following section.

Security

The subject of security related to connected vehicles was recently brought to the fore by an article published in *Wired* magazine, describing how researchers pretending to be hackers were able to gain remote control of the vehicle systems within a popular brand of vehicles [6]. The publicity associated with the incident caused the manufacturer in question to issue a recall for 1.4 million vehicles. We live in an age of ubiquitous access to information. We have the ability to remain connected to the web through wireline and wireless technologies 24 hours per day. The use of telecommunication technologies and the Internet is shared across the globe, enabling it to be a cost-effective communication means.

This sharing also means that it is open to abuse. There have been numerous high-profile incidents over the past few years in which sensitive data has

Figure 4.4 DSRC at an intersection.

been accessed by unauthorized users. There is also a perception that wireless communication links are not as secure as wireline. This raises concerns regarding the use of wireless communications for connected and autonomous vehicles. If managed appropriately, however, wireless communications can be just as secure as wireline [7].

The caveat is "if it is managed appropriately." Encryption and passwords can be used to make a wireless communication link secure, and it seems to me that people regularly trade security for convenience. It can be convenient to not have to enter a password, or to use one so weak that it is easily remembered—but such weakness also eases the path for hackers.

The use of strong passwords and encryption is key to the appropriate management of wireless communication links to ensure security. It is also important to balance cybersecurity with physical security. If a savvy criminal can talk you out of your password information, then the strength of the security is irrelevant. Also, if an unscrupulous vehicle technician has direct access to your connected autonomous vehicle, the security of your wireless communication link may be a moot point.

A balanced approach to security, requiring consideration of both cyber and physical security, enables a wireless communication link to be suitably secure while providing ease of use to the driver. The most secure system in the world would not allow the user to access it!

Driver Education

At a congestion pricing symposium just outside of London, a few years ago, a psychologist from the transport research laboratory noted the following:

> If you drive by habit you're immune to new information.

As vehicle systems become more complicated and infrastructure becomes more flexible with dynamically allocated lanes, part-time hard shoulder running, variable speed limits, and dynamic routing, there may be a need to consider driver education. It would seem reasonable that as the information technology content of the vehicle and transportation infrastructure increase that more complete training may have to be provided to the user.

Another way to address this particular challenge would be to ensure that user interfaces are well enough designed to support the needs of the average driver. A well-defined user interface allows the user to focus on the content and the task at hand, rather than trying to understand the underlying technology. When watching TV, the focus is on the content and not on understanding how TV works.

At a telematics conference in 2014 [8], it was stated that, at that time, there were more than 173 software applications available for use in the vehicle.

This suggests that there is a degree of fragmentation and a lack of standardization. Standardization is required at both the technical level and at the usability level. Focus groups have indicated that end users are struggling with the complexity and variability of the graphical user interfaces that these applications use. There are also concerns about safety related to driver distraction due to the complexity of the applications.

In the meantime, there is a growing awareness that there could be a generation gap between younger technology-savvy, less affluent connected vehicle users and older technology-averse affluent users. In addition, there is a sense that recent revelations regarding government spying activities are making consumers wary of sharing data or being tracked. On the positive side, the use of cloud-based services enables a high degree of portability between vehicles and supports situations such as stolen vehicles or vehicle changes (purchases and sales).

The relationship between the smart phone and the vehicle is also of great interest since it is recognized that the smart phone has a much larger user base than the vehicle. It also presents a more supportive platform for new technology. This could be a potential area of struggle between automotive and information technology suppliers—or an opportunity for cooperation. There is no doubt that automotive manufacturers view the connected vehicle as a prime opportunity to maintain a long-term relationship with their customers.

Data Ownership

Data ownership could be perceived as a trivial issue, but even the most advanced analytics require data. The big question is who owns the data? Clarification of the ownership of data and the existence of agreements to share the necessary data are prerequisites to the use of big data and analytics techniques. The answer to the question "who owns the data?" could be one or more of the following:

- The driver;
- The automobile manufacturer;
- The electronic supplier;
- The cloud services operator;
- The road operator.

The driver obviously has a claim on ownership as he or she has paid for the automobile and the onboard equipment. However, either the automobile manufacturer or the electronic supplier could be in a position of power, since they have control over the installation of the on-vehicle devices and potentially the data flowing to and from from the in-vehicle devices to the back-office infrastructure.

The cloud services provider may also have a role as all data must pass through the cloud from the vehicle to its ultimate destination and in reverse. The question of data ownership is also of concern to public agencies, as the real potential for probe vehicle data from connected vehicles cannot be realized if the data is not available. In the short term, the public sector should prepare to engage the private sector with a view toward data exchange.

The private sector obviously has access to a great deal of data that would be of benefit to the public sector. At the same time, due to the desire to have contextual data such as width, height, and speed restrictions, the private sector could be interested in receiving data from the public sector. A successful negotiation of this two-way data exchange will require that the public sector prepares an effective negotiating position. Public-sector data needs to be summarized and accessible. Suitable marketing materials that describe the data and the value of the data need to be prepared. Perhaps now is a good time to open discussions between the public sector and the private automotive sector regarding data sharing and cooperation?

Choosing the Best Telecommunications Approach

As discussed earlier the best telecommunications approach depends on what one is trying to do. According to the U.S. DOT, the services listed in Figure 4.5 can be supported with connected vehicles [9].

The items in Figure 4.5 might be considered a menu for the ultimate connected vehicle scenario. Many of the applications require high-speed, low-latency, reliable communications between vehicle and roadside and vehicle to vehicle of the type that only DSRC can currently provide, making that the best choice for these applications. However, the automotive industry has already moved ahead with some early market services that don't require the kind of high speed and low latency of DSRC and can thus be supported by cellular wireless and cloud-based services. These include the services depicted in Figure 4.6 [10].

Figure 4.6 provides a comprehensive menu of services that can be supported by connected vehicle technologies. Such services include navigation, entertainment, vehicle management, safety, and engine management. An important point is that the range of services is wide and will influence many aspects of vehicle operation and driver behavior. While it is likely that the private sector business model and services described above will prevail in the short term, it is also likely that later-market safety applications will require cooperation with roadside DSRC equipment and, furthermore, a more significant role from the public sector. Therefore, it would be prudent to prepare for both approaches. In the longer term, the public sector should assume that roadside equipment will

Figure 4.5   DSRC supported services.

be required for certain safety applications. Preparations should continue for the installation of roadside equipment and the coordination of actions required by the public sector in the development and deployment of infrastructure and by the private sector of the associated in-vehicle systems. It should be assumed that the private sector will lead the early market for information services and data.

However, it can also be assumed that the role of the public sector will grow as the market matures, with higher market penetrations providing support for safety applications such as collision avoidance. It should also be noted that the emergence of the autonomous vehicle with its ability to navigate roads and highways with no input from the driver—developed entirely by the private sector—could disrupt the later market for safety applications for the connected vehicle.

Figure 4.7 depicts a possible roadmap involving the early use of cloud-based services followed by DSRC-enabled services in the later market. Starting at the bottom, Figure 4.7 depicts an incremental approach that begins with the rollout of cloud-based or wide-area wireless services in the first phase. Cloud-based services addressing data collection, vehicle-to-vehicle communications,

Navigation	Entertainment	Remote Applications	Electric Vehicle	Safety
• Get directions, maps in real time. • Mobile Internet: Points of interest, broader info on the move. • Congestion/accident alerts/re-routing advice, weather/road condition alerts. • Preferred routes within city /parking guidance. • Journey planner.	• Access/play music, videos/movies/TV, games. • Internet radio. • Social networking, chat. • Trip information, schedules. • Seat adjustments. • Personalization. • Internet services.	• Remote door lock/ unlock. • Remote appliance management. • Car tracking. • Theft alerts. • Geo-fencing.	• Nearest charging station with tariff information. • Scheduling a charging slot. • Home charging. • Estimate driving range, battery charge status. • Carbon footprint.	• Speed, distance advice. • Lateral collision warning, cooperative lane change, merging assistance. • Traffic sign violation warnings. • Car breakdown warning. • Automatic call for assistance in the event of a crash (i.e., eCall), breakdown rescue. • Integrated car safety.

Vehicle Management	Fleet Management	OEM	Dealer, Service Center	Industries
• Maintenance notification alerts to individuals, garages. • Remote diagnostics. • bCall (breakdown call). • Driver performance analysis.	• Tracking and tracing, delivery notifications. • Optimal routing and journey management. • Performance analysis. • Trip records. • Virtual trainer. • Alerts and reports. • Fuel/energy management. • Resting time violation.	• Spare parts, logistics management. • Field analysis and issue identification. • Vehicle lifecycle management. • Brand improvement. • Integration with business systems. • Customer profiling.	• Warranty, logistics, inventory. • Service appointment, check-in. • Mobile workforce. • Promotions/product demos. • Product info/ guidelines. • Lifestyle profiling.	• Insurance (pay per use). • Retail: Online shopping, alerts. • Advance toll applications. • Intelligent transport systems. • Telemedicine. • Banking transactions.

Figure 4.6 Cloud-based connected vehicle services.

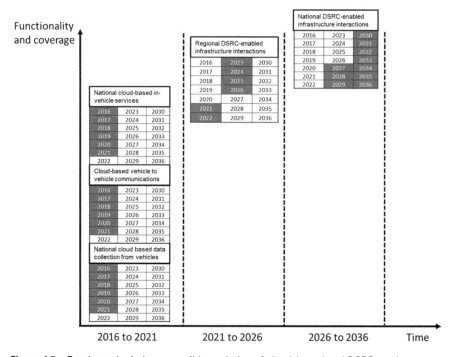

Figure 4.7 Roadmap depicting a possible evolution of cloud-based and DSRC services.

and then in-vehicle services are assumed to rollout between 2016 and 2021 as depicted by the dark shaded areas in Figure 4.7. This is followed by regional DSRC infrastructure implementation, which in turn is followed by national DSRC infrastructure implementation. This is assumed to rollout from 2021 to 2026 as indicated by Figure 4.7's dark shading. The goal is a completely connected vehicle fleet across the entire nation. This is achieved from 2027 until 2036 as represented by the dark shaded areas in Figure 4.7. DSRC rollout is assumed to come later in the evolution because of the need to install DSRC transceivers at every major intersection in the United States. The timelines represent my best estimate of how the rollout will proceed based on discussions with clients, participation in conferences, and a general assessment of the market.

Approaches to the challenge should also take account of the need to select technology approaches that are coordinated to the life cycle of the device to which they are to be installed. Typical life cycles are listed as follows:

- Road: 30–50 years;
- Vehicle: 10–15 years;
- Smart phone: 9–18 months.

These statistics make it clear that the smart phone is a more suitable environment for emerging technology as its life cycle can support accelerated deployment of new technology.

4.6   What Is an Autonomous Vehicle?

An autonomous vehicle is one that is capable of operation without a driver. Back at the turn of the century (twentieth to twenty-first, that is) this technology was referred to as an automated highway system, and significant demonstrations of it took place in both the United States and Europe. At the time, it was considered that the best way to achieve driverless operation was the use of magnetic markers inserted into the road surface, although research had also started into the use of video and light detection and ranging (LIDAR) sensors to enable vehicles to operate independently of the road infrastructure.

Today there are cars, trucks, and buses capable of completely autonomous operation using a combination of onboard sensors and computers. One such vehicle was demonstrated at the Intelligent Transportation Systems World Congress in Bordeaux, France, in 2015. The vehicle in question was a shuttle designed to carry 12 people at a time with no need for a driver. Figure 4.8 shows a photograph of the vehicle in operation taken in San Sebastian, Spain.

In addition to shuttle vehicles like the one shown in Figure 4.8, there are autonomous cars from almost every major automobile manufacturer and also from Google. Figure 4.9 shows a photograph of the Google autonomous vehicle.

Figure 4.8   EasyMile automated shuttle in operation in San Sebastian, Spain.

Figure 4.9   Google autonomous vehicle.

Autonomous vehicles are in various stages of testing, with most manufacturers agreeing that they will be on the market within the next five to eight
years. Autonomous technology can also be applied to freight vehicles. Figure
4.10 shows a road train of Volvo trucks that made use of autonomous vehicle
technology to travel across Europe.

As part of a 2016 European truck platooning challenge organized by the
Netherlands and showcased at the intertraffic exhibition in Amsterdam, automated trucks from several manufacturers, including DAF, Daimler, Iveco,

Figure 4.10   Autonomous truck platoon, Amsterdam.

MAN, Scania, and Volvo, used state-of-the art technology to drive in platoons on public roads, from various European cities to the Netherlands. They traveled on the main EU ITS corridors like the Nordic Logistic Corridor and the route between Rotterdam, Frankfurt, and Vienna. Truck platooning involves two or more trucks traveling in convoy in close proximity.

The first truck had a driver while the following trucks in the platoon were connected to this truck using wireless communications. Autonomous trucks have driven millions of miles in both the United States and Europe, and the technology is available to operate the vehicle in an autonomous mode.

The emergence of autonomous vehicles has been driven by the availability of low-cost video cameras, LIDAR, and other sensors and actuators that enable autonomous operation. This has been paralleled by the growth in the development of suitable software and the emergence of artificial intelligence and machine learning techniques. The latter can be adapted to teach vehicles how to drive without a driver.

Substantial challenges must still be overcome before the widespread A particular challenge is the definition of suitable transition arrangements that enable autonomous vehicles to be introduced on a gradual basis within a mixed traffic flow situation. In this context, some vehicles are autonomous, while some vehicles are still operated by drivers. This is particularly problematic since not all vehicles will be under automated control, and drivers could be tempted to game the situation by exhibiting dangerous driving behavior in the knowledge that the autonomous vehicles will compensate accordingly.

What is not yet clear is how both the public and transportation regulators will react to this technology. There are also some challenges associated with the gradual introduction of autonomous vehicles into traffic streams comprised of both autonomous and nonautonomous vehicles.

4.7   Autonomous Vehicle Challenges

There is a range of challenges associated with the introduction and operation of autonomous vehicles. These are explained in the following section.

Regulation

With the introduction of any new technology, a range of challenges, some of which relate to the technology itself and some of which relate to how people react to the technology, accompany the technology. If the technology is radically different, as is the case with connected and autonomous vehicles, then considerable effort may need to be made to ensure that the technology fits within existing legal and regulatory frameworks. Otherwise, new legislation and new regulations will be required. Before exploring some of the challenges associated with connected and autonomous vehicles, it is useful to consider how a new technology introduction has been handled in the past. What better example than the automobile?

The automobile was invented in 1886 when Karl Friedrich Benz invented the first true automobile powered by an internal combustion engine, fueled by gasoline. Prior to that, the use of steam-powered road locomotives in the United Kingdom, led to the introduction of a series of locomotive acts, the most notable being the Locomotive Act of 1865 [11], also referred to as the Red Flag Act. This act stipulated that road locomotives should comply with a general speed limit of 4 mph and 2 mph in towns. At the time, road locomotives were capable of speeds of up to 10 mph. The act also decreed that if the road locomotive was attached to two or more vehicles, a man with a red flag walking at least 60 yards ahead of each vehicle was mandatory. This rather draconian regulation was amended in subsequent legislation, which led to a relaxation of the speed limit to 12 mph, with the requirement for the red flag also removed.

There was a perception at the time that new technology represented a safety threat. It may also have been the case that vested interests related to horse-drawn carriages and the competing U.K. railway industry had a say in the legislation. A lesson to be learned here is that the economic benefits of the new technology (steam-driven road locomotives) were such that safety concerns and regulatory needs were quickly and effectively dealt with. The introduction of the internal combustion–driven car added to this momentum for progress.

The past echoes as we react to self-driving vehicles in the way we reacted to stream-driven road locomotives in the past. It provides the basis for great optimism for future policymaking, as history shows that market forces will eventually overcome regulatory issues.

Getting There in Stages

With respect to the connected vehicle there are very few issues when it comes to transitioning from the current situation to the fully equipped connected vehicle. As discussed in Section 4.4, there are two distinct approaches but in any event it is possible that the technology is there and that there are few barriers to implementation. With respect to the autonomous vehicle, we have a different situation. One of the challenges is the gradual and safe introduction of autonomous vehicles, moving from the current situation where every vehicle has a driver, through some level of partial automation, to complete automation. In addressing terminology associated with the autonomous vehicle, the Society of Automotive Engineers has developed a framework that explains the probable transition from today to tomorrow. This is illustrated in Figure 4.11.

SAE level	Name	Narrative Definition	Execution of Steering and Acceleration/ Deceleration	Monitoring of Driving Environment	Fallback Performance of *Dynamic Driving Task*	System Capability (*Driving Modes*)
Human driver monitors the driving environment						
0	No Automation	the full-time performance by the *human driver* of all aspects of the *dynamic driving task*, even when enhanced by warning or intervention systems	Human driver	Human driver	Human driver	n/a
1	Driver Assistance	the *driving mode*-specific execution by a driver assistance system of either steering or acceleration/deceleration using information about the driving environment and with the expectation that the *human driver* perform all remaining aspects of the *dynamic driving task*	Human driver and system	Human driver	Human driver	Some driving modes
2	Partial Automation	the *driving mode*-specific execution by one or more driver assistance systems of both steering and acceleration/ deceleration using information about the driving environment and with the expectation that the *human driver* perform all remaining aspects of the *dynamic driving task*	System	Human driver	Human driver	Some driving modes
Automated driving system ("system") monitors the driving environment						
3	Conditional Automation	the *driving mode*-specific performance by an *automated driving system* of all aspects of the dynamic driving task with the expectation that the *human driver* will respond appropriately to a *request to intervene*	System	System	Human driver	Some driving modes
4	High Automation	the *driving mode*-specific performance by an automated driving system of all aspects of the *dynamic driving task*, even if a *human driver* does not respond appropriately to a *request to intervene*	System	System	System	Some driving modes
5	Full Automation	the full-time performance by an *automated driving system* of all aspects of the *dynamic driving task* under all roadway and environmental conditions that can be managed by a *human driver*	System	System	System	All driving modes

Figure 4.11 The Society of Automotive Engineers' autonomous vehicle evolution table.

4.8 Summary of the Differences between Connected and Autonomous Vehicles

It is valuable at this point to provide a summary of the differences between connected and autonomous vehicles. Connected and autonomous vehicles have considerable overlap in the communities that are conducting the research and development. Often conferences and meetings are held under the banner of the combined subject. While both involve the application of advanced technologies to transportation, it is important to distinguish between connected vehicles and autonomous vehicles. Figures 4.12 and 4.13 attempt to achieve this in a graphical manner.

Figure 4.12 illustrates that the connected car consists of a car with the addition of the ability to communicate with the outside world using either roadside DSRC technology or cellular technology, making use of cloud services. Of course the connected vehicle also includes vehicle-to vehicle (V2V) communications as depicted in Figure 4.13.

The autonomous vehicle is one that does not require a driver, as depicted in Figure 4.14.

This illustrates the point that a car with the driver removed can be considered as an autonomous vehicle. Of course, the same can be true for buses and trucks. It could be argued that autonomous operation is more valuable for buses and trucks because the driver is not attempting to get anywhere and has to be present in the vehicle as a requirement for operation. Substantial reductions in cost for transit and freight could be achieved by the introduction of autonomous or driverless operation. The other major difference between connected and autonomous vehicles is that connected vehicles are highly likely to appear in the marketplace in substantial numbers before Autonomous vehicles.

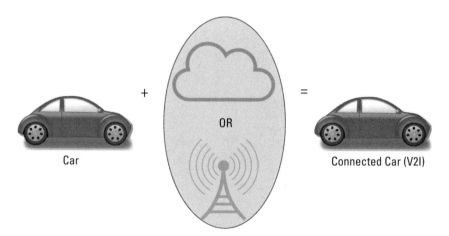

Figure 4.12   The connected vehicle.

Figure 4.13   Vehicle-to-vehicle communications between connected vehicles.

Figure 4.14   The autonomous car.

4.8   Connected and Autonomous Vehicles within a Smart City

The smart city will provide a wider technology context within which connected and autonomous vehicles will operate. One aspect of the smart city will be the use of the internet of things (IoT). As connected and the autonomous vehicles have emerged, interest has grown substantially in the IoT. It is viewed as the next generation of the Internet, and it is predicted that it will go beyond connecting computers and smart phones to connecting a multitude of different devices including refrigerators, air conditioning systems, homes, offices, retail systems, and financial systems.

The notion is that the connected vehicle will reach beyond the vehicle, connecting to the back office, and will ultimately support connections between the vehicle and many other things. Google's recent acquisition of Nest and Apple's announcement of an entry into home automation support the trend toward the IoT. It also seems that the IoT is being used as a marketing tool to explain to drivers the values and benefits of the connected vehicle. An important message of the IoT, at least from a transportation perspective, is that the vehicle is not the center of the universe, but part of a larger network of things that are connected and services that can be provided to people. Dan Teeter, the director of vehicle connected services for Nissan, summed this up with the following description of a typical day in the life of an IoT citizen [8]:

> A driver named Pat drives to work in his autonomous car. His smart watch and seat belt sensor detect that his blood pressure and heart rate are elevated, so the car switches to soothing music and a back rub. Arriving at work, the car drops Pat off and proceeds on its own to a prepaid parking

spot. It asks the home refrigerator to check for healthy food, and finding none, it places an order with a local health food store. When Pat decides to clock out early, the car alerts a connected thermostat to start cooling his house earlier than usual. The car figures out the best route home. It automatically pays tolls along the way. And it plays a comedy movie on Pat's screen to help him relax.

This picture of a potential future for the connected vehicle provides a good illustration of the point that the private sector is considering services that are significantly beyond a simple connection from the vehicle to a central database within a back office. It also provides an indication of how the connected vehicle and the autonomous vehicle will integrate within a wider framework such as a smart city.

With respect to autonomous vehicles, it is likely that these will be used to enable on-demand transportation services using driverless vehicles. Uber is already experimenting with such an approach, and it is not too difficult to envision the future when the Uber service is provided by autonomous vehicles. Smart cities are discussed in detail in Chapter 5.

4.9   The Likely Impact of the Connected and the Autonomous Vehicle on Transportation

While it is likely that the private sector business models described above will prevail in the short term, it is also likely that later-market safety applications will require cooperation with roadside equipment and feature a more significant role from the public sector. Therefore, it would be prudent to prepare for both models. In the short term, the public sector should prepare to engage the private sector with a view toward data exchange.

The private sector obviously has access to a great deal of data that would be beneficial to the public sector. There is an indication, through the private sector's desire to have contextual data such as road geometry, speed limits, and height/weight restrictions, that it could be interested in receiving data from the public sector. A successful negotiation of this two-way data exchange will require that the public sector prepare an effective negotiating position. Public sector data needs to be summarized and accessible. Suitable marketing materials that describe the data and the value of the data need to be prepared.

In the longer term, the public sector should assume that roadside equipment will be required for certain safety applications. Therefore, preparations should continue for the installation of roadside equipment and the coordination of actions required by the public sector for infrastructure-based development and by the private sector with respect to the connected vehicle. It should be assumed that the private sector will lead the early market for information services and data. However, it can also be assumed that the role of the public sector will grow as the market matures when higher market penetrations will provide support for safety applications such as collision avoidance. It should also be noted that the emergence of autonomous vehicles, with their ability to navigate roads and highways with no input from the driver and developed entirely by the private sector, could disrupt the later market for safety applications.

4.10 Big Data and Connectivity

An important aspect of the use of big data that is impacted by connected vehicles lies in the relative cost of vehicle probe data from connected vehicles compared to the collection of data from fixed roadside infrastructure sensors. The growing availability of private sector data from the vehicle is likely to place the vehicle at the center of the data collection universe. In order to understand how this should influence future data collection and data acquisition strategies for the public sector, it will be necessary to assess the relative cost of vehicle probe data versus fixed-sensor data. The cost of installation, maintenance, sensor operation, and data management should be compared against the cost of acquiring data from connected vehicle operators. This latter option is like that of current private data operators such as INRIX [12] and HERE [13]. Both connected and autonomous vehicles will benefit from the application of big data, converging with the two-way conductivity to and from the vehicle. Possibilities will emerge for automated fleet operation and a higher level of decision support for vehicles that still have human operators. The intelligence of automated vehicles will also improve using big data and conductivity. Analytics conducted on a big data set will reveal trends and patterns in both vehicle operation and driver behavior that can be incorporated into future artificial intelligence and machine learning approaches. Conductivity will support the extraction of data from vehicles and the provision of control and decision-support information to the vehicles.

4.11 Connected and Autonomous Vehicles within a Smart City

It is obvious that an important element of a smart city from a transportation perspective will be the use of autonomous vehicles. These will include private cars, transit vehicles, and freight vehicles that are able to operate without a driver. It is worth looking ahead to what the possible shape of a smart city that incorporates autonomous vehicles.

A possible scenario for autonomous vehicles would be in support of on-demand transportation services. This would reduce the need to own a vehicle and extend the concept of as a service from information technology to transportation. The concept could emerge as driverless Uber or as an adaptation of the current rental car approach, with rental car companies operating a fleet of autonomous vehicles to provide on-demand transportation. A subscriber would be able to use a smart phone app to summon a vehicle on demand, and sophisticated algorithms (probably already in place at Uber) would allocate the most appropriate vehicle and dispatch it to the subscriber. Under this scenario, it is likely that there would be more miles traveled by a lesser number of vehicles. The basis for this thinking is that autonomous vehicles would operate 24 hours a day, seven days a week, generating more miles than current vehicles. This would obviate the need for people to own their own vehicles, hence reducing the number of vehicles on the road.

While it is also possible that individuals will acquire their own autonomous vehicles, perhaps this fleet approach represents a useful transition strategy from manually operated to autonomous vehicles. The operation of an autonomous fleet by professional managers and operators would help with the smooth introduction of the technology. Fleet operation of autonomous vehicles could also have significant impact on transit services and freight delivery services within a smart city. There is also considerable scope for applying big data and data analytics to the needs of the transportation disadvantaged. The efficiency and effectiveness of on-demand transportation services for those people who cannot use conventional route services could be greatly improved by the use of such techniques.

There is a convergence between the operation of an autonomous vehicle fleet and these needs. While considerable publicity has been generated by the notion that freight in urban areas could be delivered by flying drones, it is more likely that freight will be delivered by driving drones. Of course, all this will be taking place within the context of a smart city where technology has been harnessed to improve the quality of life for citizens and visitors.

Looking at this wider context, one can imagine that from a transportation point of view, a smart city would offer a range of complementary services supported by smart phone apps, a sophisticated communication network, and processing and analytics capability. Smart city services will be driven by a deeper understanding of the demand for transportation and current operating conditions. This will be enabled by a richer stream of data and by more sophisticated analytics. Infrastructure-based sensors would be an integral part of the smart city, but equally important would be the data-generation possibilities from autonomous vehicles and connected vehicles. It is not difficult to imagine a smart city the delivers a range of useful services to both citizens and visitors based on this new understanding of what's going on and what the needs are any given time.

This also leads to the idea that a really smart city would be aware of what's going on in its surroundings as well as within the city boundaries. The sensing capabilities of a smart city should extend to the original origin of visitors to the city and be able to understand where they're coming from, why they are traveling, and what they do when they get to the smart city.

As is the case with all emerging technologies, there is a temptation to identify a solution and then look for a problem. So the really smart city would apply services based on identified and defined needs, issues, problems, and objectives.

The Emerging Technologies Forum of ITS America [14] is working to provide support materials on this subject. A group of people drawn from both the public and private sectors is defining a checklist of services that would be expected in a smart city and a catalog of use cases that can be incorporated into smart city plans. A use case is simply a description of a problem that can be addressed and how it can be done.

It is likely that autonomous vehicles have an extremely important role to play in the future smart city and that the introduction of this technology will have a dramatic effect both on how we use transportation and, ultimately, on urban land use. It is likely that progress with autonomous vehicles will be mirrored by progress in back-office automation, as our understanding of demand and operating conditions enables a higher proportion of automated event triggering. The smart city of tomorrow will make use of an internet of transportation to provide a higher level of decision support to transportation operators. Perhaps we will reflect on the current situation in 50 years' time and wonder what the transportation profession and automotive manufacturers were thinking in allowing people to actually drive vehicles?

4.12   The Likely Effect of Connected and Autonomous Vehicles on the Automotive Industry

Gartner predicts, "by 2020, there will be a quarter of a billion connected vehicles on the road" [15], enabling new in-vehicle services and automated driving capabilities There is a general sense that the future direction of the automotive industry is becoming clearer and that technology will define the future. As we discussed at the beginning of the chapter, the electronics content of the automobile has been increasing steadily since the 1950s, which has set the scene for a much more active involvement of the IT industry in the traditional automotive business.

This is interesting since typically the automotive industry has focused on developing and providing robust technology, well-proven products that are reliable enough to avoid large-scale recalls. Automotive manufacturers clearly understand that the average vehicle will stay on the road in the U.S. fleet for an average of 11.5 years [16] and that any technology advances have to be sustainable enough to endure for that length of time. However, there are a number of new players (and old players) offering solutions that make use of information technology to link the driver to a range of services and to link the driver to the automotive manufacturer.

It is also interesting to note the deep interest from insurance companies and actuaries focusing on using connected vehicle data to develop more realistic insurance rates based on vehicle position and driver behavior. Some people also believe that the connected vehicle can act as a stepping stone to the autonomous vehicle and consider it the job of telematics to cover both connected and autonomous vehicles. It could also be imagined that insurance companies' interest in the connected vehicle is in part driven by the desire to understand driver and vehicle behavior as the autonomous vehicle emerges.

In the meantime, some in the automotive industry are perplexed by the idea of technology shaping the future of the automotive industry, due to their concern that information technology companies will eventually lead the automotive market. This has been compounded by Google's recent announcement that it will develop its own autonomous vehicle and by the presence of both Apple and Google in the connected vehicle market. In particular, Google's intentions seem to be regarded with a degree of suspicion. However, considering the discrepancy between how much time we spend in the car and how much money is spent on in-car advertising, I view Google primarily as an advertising company and its entry into this market as nothing more than a logical extension of that activity.

One of the most interesting aspects of the current connected vehicle market is the number of participants who have emerged in the middle ground between the driver and the automotive manufacturer. Some of these participants are relatively new while others have been in the business for 10–15 years and support early initiatives such as General Motors' OnStar service [17]. These players are information technology, consumer electronics, and telecommunications companies providing cloud-based services to the automotive manufacturers.

They include the following companies:

- Wirelesscar;
- Airbiquity;
- Covisint;
- Racowireless;
- Sprint Velocity;
- Verizon.

An indication of the extent to which the private sector has developed and embraced the connected vehicle was found in some statistics from Airbiquity.

They show that Airbiquity had more than six million connected subscriptions in 2015. Over three billion connected card transactions have been processed to date, averaging 250 million transactions per month worldwide. They are supporting more than 100 million transactions per month [18].

4.13 Summary

It is obvious that the connected vehicle, at least in the eyes of the private sector, now exists. The main focus of the private sector is now on how to monetize the connected vehicle. It is not yet clear whether this initiative will be driven by consumer electronics and information technology companies, or by automotive manufacturers. While there is some variation and a lack of standardization in many of the approaches, it should be assumed that the connected vehicle is here to stay and that public perception and regulations will fall into line with market forces. With respect to public sector strategies and activities, it is necessary to recognize the existence and extent of connected vehicle initiatives in the private sector. It is also prudent to engage some of the key private sector players to provide further understanding of their business directions and motivations and explore the possibilities for two-way data information exchange agreements.

The connected vehicle is a major focus for both federal and private sector investment. The ability to establish a reliable two-way communication channel between drivers and an information technology infrastructure holds the promise of substantial advances in safety, efficiency, and user experience. This equates to advances such as a reduction in crashes, more reliable trip times, and more effective traveler information.

For example, data regarding the current operating conditions of the vehicle, the vehicle's current location, and the vehicle ID, along with information on driver conditions, could be transmitted to a back office. Vehicles that experience a thunderstorm coming in one direction along the highway could warn vehicles coming the other way that they are about to encounter the storm. It is also possible that two vehicles could negotiate a mutually agreed on exclusion zone around each other, thus preventing collisions and conflicts. At traffic-signalized intersections, cooperation between connected vehicles and the traffic signal system could mean an end to red light running and provide warnings to drivers that vehicles on other approaches to the intersection may be about to violate a red light.

With respect to the autonomous vehicle, just like the steam road locomotives of the 1860s, it can be assumed that the initial concerns regarding public perception and the need for new regulation will be overcome by market forces. The possibility that on-demand services can be supplied using autonomous vehicle fleets holds the promise of urban transportation being revolutionized and, further, may facilitate the provision of mobility as a service.

References

[1]   https://www.pwc.com/gx/en/technology/publications/assets/pwc-semiconductor-survey-interactive.pdf, retrieved August 12, 2016.

[2]   Wikipedia, https://en.wikipedia.org/wiki/Automotive_electronics, retrieved August 12, 2016.

[3]   http://asmarterplanet.com/blog/2010/05/a-systems-approach-to-rethinking-transportation-from-the-itsa-annual-conference.html, retrieved August 12, 2016.

[4]   https://www.airbiquity.com/choreo-platform/ retrieved August 15, 2016.

[5]   http://www.its.dot.gov/communications/image_gallery/image19.htm, retrieved September 5, 2016.

[6]   *Wired* magazine article https://www.wired.com/2015/07/hackers-remotely-kill-jeep-highway/, retrieved September 7, 2016.

[7]   http://blogs.cisco.com/home/wireless_networks_vs__wired_which_network_is_more_secure, retrieved August 15, 2016.

[8]   http://analysis.telematicsupdate.com/infotainment/telematics-detroit-2014-day-one, retrieved August 15, 2016.

[9]   http://www.its.dot.gov/pilots/cv_pilot_apps.htm, retrieved September 5, 2016.

[10]   https://www.cognizant.com/InsightsWhitepapers/Exploring-the-Connected-Car.pdf, retrieved September 5, 2016.

[11]   https://en.wikipedia.org/wiki/Locomotive_Acts, retrieved September 5, 2016.

[12]   http://inrix.com/, retrieved September 5, 2016.

[13]   http://mapupdate.navigation.com/landing/en-US/?gclid=COH3jbHK-c4CFdgKgQodfEIMDQ&gclsrc=aw.ds, retrieved September 5, 2016.

[14]   http://www.itsa.org/, retrieved September 5, 2016 at 9:27.

[15]   http://www.gartner.com/newsroom/id/2970017, retrieved September 5, 2016.

[16]   http://www.usatoday.com/story/money/2015/07/29/new-car-sales-soaring-but-cars-getting-older-too/30821191/, retrieved September 5, 2016.

[17]   https://www.onstar.com/us/en/home.html, retrieved September 5, 2016.

[18]   https://www.airbiquity.com/news/press-releases/airbiquity-enters-second-half-2015-surpassing-six-million-connected-cars-its-choreo-cloud-platform/, retrieved August 15, 2016.

5

Smart Cities

5.1 Informational Objectives

This chapter answers the following questions:

- What is a smart city?
- What objectives can be addressed by a smart city?
- What steps can be taken toward implementing a smart city?
- How can smart city investment activities be coordinated?
- How can the effects of smart city investments be evaluated?
- What challenges and opportunities are associated with smart cities?
- What Is the Sentient City concept?

5.2 Chapter Word Cloud

As provided in previous chapters, Figure 5.1 shows a word cloud for this chapter, with fonts for each word proportional to the frequency of use of that word.

5.3 Introduction

The term *smart cities* has become increasingly popular over the past 10 years. Figure 5.2 shows growth in the interest in smart cities (dark gray) compared to

Figure 5.1 Word cloud for Chapter 5.

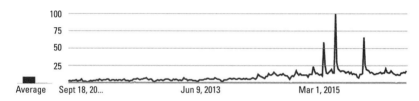

Figure 5.2 Global interest in smart cities compared to intelligent transportation systems, 2006–2016.

interest in intelligent transportation systems (light gray), as measured by the number of Google inquiries [1].

Urban population growth, coupled with a developing awareness of the value of technology in addressing urban problems, has driven interest in the subject. Smart cities has emerged as an umbrella label for the application of advanced technologies to the needs, issues, problems, and objectives of people living within the urban environment. This chapter provides an overview of the entire smart city spectrum of technology applications and then details transportation, mobility, and accessibility approaches to smart cities.

Smart city initiatives are occurring within the context of significant prior deployment of intelligent transportation systems and services over the past 20 years. In more recent times, this has also included research and pilot deployment of connected and autonomous vehicles. Along with the opportunities associated with the smart city comes the challenge of integrating this prior work with new initiatives. A smart city provides a wider context for the application of advanced technologies to transportation and brings a range of new partners with different perspectives. Successful planning, design, and implementation of smart cities will require that the transportation profession reaches out to these new partners ensuring that the transportation elements of the smart city are seamlessly connected to other elements.

5.4 What Is a Smart City?

A search for a definitive definition of the smart city proved inconclusive. It is early in the development and application of smart city technologies, so perhaps convergence on a single definition will occur in the future. In the meantime, the Emerging Technologies Forum of ITS America has decided to adopt a working definition of a smart city to guide its work on the subject. That working definition is the following:

> You know your city is smart if you poke with a stick and it reacts appropriately.

Although light-hearted, this definition captures the overall essence of a smart city at the highest level. It requires that a smart city can sense opportunities, threats, and changes within the city and the wider context. It also assumes that a smart city has sufficient intelligence to be able to develop an appropriate response. In a more detailed look at the definition of a smart city included in a recent White House report [2], smart city infrastructure can be summarized as shown in Figure 5.3.

Note that Figure 5.3's definition of a smart city includes energy, smart buildings, utilities, manufacturing, and agriculture as well as transportation. It is important to note that a smart city involves the application of advanced technologies to a wide range of services and that transportation represents a subset of such services.

According to the U.S. Department of Transportation [3], which naturally takes a transportation-centric view of the smart city, the smart city consists of several vision elements, as shown in Figure 5.4 and described in the following sections.

Urban Automation

Urban automation includes driverless private vehicles, freight, logistics, and transit vehicles. It could also cover the use of drones to make deliveries.

Connected Vehicles

The U.S. DOT envisions that connected vehicles (discussed in Chapter 4) will be connected to other vehicles and to back-office infrastructure via the use of DSRC.

Urban Sector	Technologies / Concepts	Objectives
Transportation	Multi-modal integration via ICT applications and models On-demand digitally enabled transportation Design for biking and walking Electrification of motorized transportation Autonomous vehicles	Save time Comfort or productivity Low-cost mobility and universal access Reduced operating expenses to transportation providers Zero emissions, collisions, fatalities Noise reduction Lifestyles Tailored solutions for the underserved, disabled, and elderly
Energy	Distributed renewables Co-generation District heating and cooling Low-cost energy storage Smart-grids, micro-grids Energy-efficient lighting Advanced HVAC systems	Energy efficiency Zero air pollution Low noise Synergistic resource management with water and transportation Increased resilience against climate change and natural disasters
Building and Housing	New construction technologies and designs Life-course design and optimization Sensing and actuation for real-time space management Adaptive space design Financing, codes, and standards conducive to innovation	Affordable housing Healthy living and working environments Inexpensive innovation and entrepreneurial space Thermal comfort Increased resilience
Water	Integrated water systems design and management Local recycling Water efficiency via smart metering Re-use in buildings and districts	Active ecosystem integration Smart integration of water, sanitation, flood control, agriculture, and the environment as a system Increased resilience
Urban Manufacturing	High-tech, on-demand 3D printing Small batch manufacturing High-value added activities requiring human capital and design Innovation parks	New job creation Training and education Urban space conversion and re-use Close integration of living and work
Urban Farming	Urban agriculture and vertical farming	Lower water use Cleaner delivery Fresher produce

Figure 5.3   City infrastructure technologies.

Intelligent, Sensor-Based Infrastructure

This vision element entails the use of transportation infrastructure-based sensors to provide the data required for service quality management and as the basis for connected citizen and connected visitor information and transportation management.

User-Focused Mobility Services and Choices

This vision element foresees a combination of public and private mobility services including Uber and Lyft to create a portfolio of mobility services and options for the traveler.

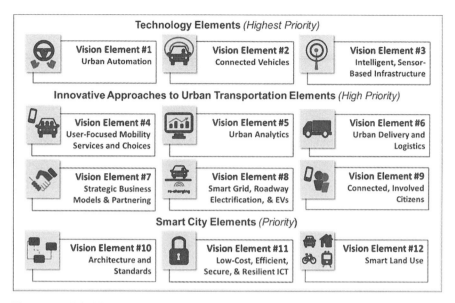

Figure 5.4   U.S. DOT smart city vision elements.

Urban Analytics

Big data and analytics will be used to understand prevailing transportation conditions and the demand for transportation and to provide the data required for insight and understanding. This also involves the provision of the analytics tools required to turn data into information then into actionable strategies.

Urban Delivery and Logistics

In this vision element, advanced technologies will be applied to optimize urban freight delivery and the logistics processes involved. This includes fleet management and advanced technology systems to improve the efficiency of freight delivery.

Strategic Business Models and Partnering

This element entails the definition, establishment, and ongoing management of clearly defined business models that identify sources of investment and assign rewards, roles, and responsibilities to both public and private sector partners.

Smart Grid, Roadway Electrification, and Electric Vehicle

Smart cities should aim to optimize power distribution and implement the infrastructure required to support the electric vehicle and the provision of energy to the electric vehicle from electrified roadway installations. This also includes the design, development, and establishment of a network of charging points for the electric vehicle.

Connected, involved citizens.

A further vision element calls for the use of smart phones and other technologies to ensure that citizens within the smart city are fully connected with information and government services.

Architectures and standards.

Smart cities must also aim to use best industry practices for technology, organizational, and business model frameworks. This includes the adoption of relevant national and international standards and the development of local standards where appropriate.

Low-cost, efficient, secure, and resilient Information and Communications Technology (ICT).

A further vision element calls for the use of appropriate communications and processing technologies, including wireless and wireline, to support data transmission and information sharing.

Smart land use.

Since there is a well-documented relationship between land use and the demand for transportation, a smart city should also include long-range plans to adapt land use to optimize transportation service delivery, accessibility, and mobility. In addition, smart cities can use urban analytics to improve their understanding of the effects of land use on transportation demand, such as the observation and analysis of trip generation from zones with a predominant land use.

Another perspective on the definition of a smart city is that of the smart cities council [4]. The council identifies three core functions of a smart city. These functions, which are shown in Figure 5.5, are described as follows.

- Collect: Data is collected from a wide range of devices and sensors, both vehicle and infrastructure-based.
- Communicate: This takes advantage of a combination of wired and wireless communications to bring the data back to a central back office for processing.
- Crunch: This is the smart cities council term for data processing, which is conducted for three purposes:
  - Presenting;
  - Perfecting;
  - Predicting.

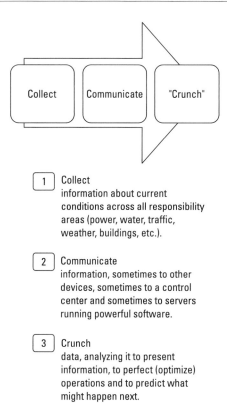

Figure 5.5 Smart cities council definition of a smart city.

During processing the data is turned into information, and actionable insights are gleaned that can form the basis for transportation service delivery strategies. It is interesting to note that this latter definition is close to the working definition adopted by the Emerging Technologies Forum of ITS America, as discussed earlier.

The service definitions described so far provide an overview of a smart city in terms of a wide range of services beyond transportation and begin to converge on the definition of a smart city from a transportation perspective. The remainder of this chapter focuses on the definition of a smart city from a transportation perspective, although we will discuss other elements of the smart city to retain context.

To converge on a detailed definition of a smart city, from a transportation perspective, we take the U.S. DOT smart city vision elements as a starting point [3] and supplement them with vision elements from each of the seven U.S. DOT smart city challenge finalist's applications [5–11] to form the basis of the following list of smart city transportation services. Note that the vision elements have been transformed into services as these deliver value in terms of safety, efficiency, and user experience, while projects and technologies simply enable the establishment and operation of the service. Section 5.6 discusses this in more detail.

Asset and Maintenance Management Services

A smart city will consist of considerable investment in infrastructure and advanced technologies that will have to be maintained, tracked, and managed beyond the initial implementation. This range of services provides the tools and decision-support necessary to manage the smart city assets and apply consistent maintenance standards in a timely manner. An investment in smart cities can be thought of as a whale-shaped investment with the head of the whale representing the initial capital investment and the whale tail representing the operations and maintenance investment. One of the challenges that have been encountered in the past with respect to the application of advanced technologies to transportation has been the allocation of resources to the whale tail. Good asset and maintenance management not only provides decision support; it also provides the information required to justify continued investment in this important area.

Connected Vehicle Services

The subject is addressed extensively in Chapter 4. In the smart city, connected vehicle technology will be applied to private cars, freight vehicles, transit vehicles, and vehicles that provide MaaS such as Uber, Lyft, taxis, and paratransit entities. A range of services will be provided to smart city citizens and visitors.

Connected, Involved Citizen Services

This group of services involves the use of smart phones and other communication technologies to establish and maintain two-way communications with smart city citizens.

Integrated Electronic Payment Services

Payment for transportation services within the smart city will be facilitated through a range of electronic payment services. These will be supported by an integrated citywide electronic payment system that addresses tolls, transit tickets, and parking fees. It is also likely that this citywide system will support payment for government services.

Intelligent, Sensor-Based Infrastructure Services

Smart sensors and the appropriate telecommunications network technology will be utilized within the smart city to provide a range of data supply services to both public and private sector entities within the smart city. Sensor-based data services will be complemented by probe vehicle data coming from connected vehicles. Sensor data will also be supplemented by mobile data from smart phones and social media data from social media networks such as Twitter. This will enable smart city managers to measure performance and gauge perception of services with respect to transportation. Data from smart sensors will also provide a significant input into back-office activities that deliver analytic services

Low-Cost Efficient, Secure, and Resilient ICT Services

Data will be transmitted within the smart city using a range of information, communication, and technology services. Some of these will be owned and operated by the public sector and others by the private sector. It is likely that a hybrid approach involving wireline and wireless solutions will be adopted to supply the services. The communication services will be delivered within a managed framework that provides maximum efficiency and security, while taking a network approach to supporting resilience.

Smart Grid, Roadway Electrification, and Electric Vehicle Services

This involves electrification of vehicles and the establishment of a network of charging points to enable electric vehicles to have the same range as gasoline-powered vehicles. This will require that electric vehicle charging points become as ubiquitous as gas stations. The services provided will be enabled using renewable energy to optimize the cost of energy and the ensuing emissions. Services will be provided via private cars, transit systems, and electric fleets.

Smart Land-Use Services

Smart land use within a smart city will include the establishment of mobility and travel access hubs that support multimodal transfer and the optimization of the daily commute. Urban analytics will also support smart land use by providing further insight into the relationship between land use and transportation demand. Smart land use will also involve detailed consideration of accessibility to jobs and the relationship between residential zones and work zones.

Strategic Business Models and Partnering Services

The services to be delivered in the smart city will be enabled by strategic business models and partnering. While previous models for deploying advanced technologies in transportation have had a predominantly public-sector focus, it is likely that the smart city will be supported by public-private partnerships that harness the resources and motivation of the private sector in addition to the public sector.

Transportation Governance Services

Transportation governance services within a smart city will ensure the optimized management of transportation on a multimodal basis across the whole city. This will require collaboration and coordination between public- and private-sector service providers and involve cross-mode cooperation for the management of private vehicles, transit vehicles, and freight vehicles. It is likely that advanced data capture and data-processing technologies will be used to support government transportation services. This also includes the application of architecture, standards, and prevailing best practices.

Transportation Management Services

Transportation management involves the coherent and consistent management of all modes of transportation within the smart city. This includes private vehicles, transit vehicles, freight vehicles, and nonmotorized means of transportation such as bicycling and walking. Essential elements of transportation management will include the application of performance management techniques to the entire transportation system. Data from various sources along with sophisticated data management techniques will be used to measure the performance of transportation services and compare them to predefined yardsticks.

Transportation management will incorporate current traffic management approaches to support a range of services including freeway incident management, emission testing and mitigation, highway and railroad intersection, automated vehicles, connected vehicles, and advanced traffic signal control for urban surface streets and arterials.

Traffic management services will include integrated corridor management, travel demand management, dynamic parking management, pedestrian mobility applications, and the use of the IoT to increase the sustainability of smart city transportation approaches. Traffic management also extends to the management of emergency vehicles, fleet deployment, and incident prediction. This will also incorporate the use of big data and analytics techniques to support a range of services that will underpin results-driven investment programs, through which future transportation investments are guided by the effects of prior investments.

Travel Information Services

Travel information will be delivered on a multichannel basis that includes in-vehicle information, smart phone information, roadside information, and the delivery of information services in the home and the office via the web. Travel information services will support the connected citizen and the connected visitor and will act as a sophisticated form of transportation management by making decision-quality information available to all citizens and visitors within the city. Travel information will offer choices regarding route, timing of journey, and mode of travel and will include information regarding total trip time, trip time reliability, and cost of various options.

Urban Analytics Services

Urban analytics within a smart city will be utilized to measure and improve access to jobs, transportation safety, and transportation efficiency and to provide the enabling services to manage transportation user experience. Urban analytics will be driven by data links, centralized data hubs, and enterprise data management systems that make use of big data techniques to provide the raw material for sophisticated analytics. The services will also be enabled by open data cloud approaches.

Urban Automation Services

Urban automation services will be supported by autonomous vehicle technology for private vehicles, driverless shuttles, company and public agency fleets, taxis, and paratransit. Urban automation will also extend to the use of driverless vehicles and drones to deliver packages.

Urban Delivery and Logistics Services

This involves the use of advanced information and communication technologies to improve and optimize delivery and logistics within the smart city environment. This will include information services to the driver and fleet managers regarding traffic congestion and the delivery of optimized routing information services.

User-Focused Mobility Services

Public and private supported mobility services will be combined into a portfolio of choices available to the smart city traveler. It is likely that a marketplace approach will be taken to this, allowing for the acquisition of a range of mobility services through a central information point or website. This range of services will be complemented by travel information services that can provide detailed information on the mobility services choices available.

5.5   Smart City Objectives

The root objectives of the smart city from a transportation perspective are safety, efficiency, and enhanced user experience. These objectives can be decomposed into a more detailed list of objectives. It is likely that smart city objectives will consist of at least two levels—higher-level objectives that capture policy and more detailed objectives that are used to guide planning and delivery. Examples of high-level policy objectives are listed as follows:

- Inducing a 1% modal shift in favor transit;
- Improving customer perception of regional transportation by 10%;
- Reducing regional traffic delays by 10%;
- Improving the reliability of trip times by 15%;
- Reducing accidents by 10%;
- Improving incident response time by 20%.

Figure 5.6 shows the more detailed smart city objectives used to guide deployment in a tabular fashion, with the objectives related to the smart city transportation services we have defined.

Note that the objectives are closely related to the 20 big questions defined in Chapter 2.

5.6   Steps Toward a Smart City

A roadmap toward a smart city will consist of a departure point, a journey, and a destination point. It is likely that the departure point for a smart city will vary considerably depending on the perception of need within any city and the pattern of prior investment. For example, if significant investment has been made into electronic toll systems or electronic transit ticketing systems, then a suitable departure point on the road toward a smart city may lie in a citywide electronic payment system that provides services for toll payment, transit ticketing, and payment of parking fees. An electronic payment departure point also provides the benefit of a means to collect revenue in the most cost-effective manner, which is always useful at the beginning of a new program. Another city could have a significant prior investment in traveler information services, making the establishment of the connected citizen and the connected visitor programs an attractive option as the point of departure.

At the other end of the journey, there lies a destination point for the smart city. It is to be hoped that all cities will eventually share a common view of the destination and converge on a vision of the ultimate smart city. As discussed earlier in Section 5.4, such convergence does not yet exist as we are very early in the development and deployment of smart cities. It is vitally important that any smart city program has a clear definition of the ultimate outcome as this will play a significant role in developing the roadmap from today's situation to the future desired situation.

With respect to the journey, it would be very useful to define a series of steps or milestones that cities can take to get from the departure to the destination point. The traditional way to define the roadmap would be to develop an investment program that consists of projects and programs. While this provides

Objectives \ Services	User focused mobility	Urban delivery and logistics	Urban automation	Urban analytics	Traveler information	Transportation management	Transportation governance	Strategic business models and partnering	Smart land use	Smart grid, roadway electrification and electric vehicle	Low cost efficient, secure and resilient ICT	Intelligent sensor-based infrastructure	Integrated electronic payment	Connected, involved citizens	Connected vehicle	Asset and maintenance management
Safety																
Maximizing the safety of the overall transportation system		●	●	●	●	●		●		●	●	●		●	●	●
Balancing safety, efficiency and user experience to create a holistic approach			●	●	●	●	●	●								
Understanding the effects of safety improvements and investments				●												
Efficiency																
Identifying bottlenecks and slowdowns in the transportation system and developing strategies to manage these			●	●	●	●					●	●	●	●	●	
Optimizing the performance of transportation assets and infrastructure			●	●				●			●	●				●
Optimizing the performance of the overall transportation system			●	●		●	●	●								
Minimizing capital and operating costs optimizing current and future expenditure		●	●	●	●	●	●	●	●		●	●	●	●	●	●
Improving service levels for citizens and visitors	●						●	●	●	●				●	●	●
Identifying and addressing service deficiencies in the transportation system				●		●	●				●	●	●			

Figure 5.6 Smart city objectives and transportation services.

Assessing current and future demand for transportation					●			●	●	●	●	●	
Optimizing access to jobs		●			●			●	●	●	●	●	●
Optimizing transportation service for the transportation disadvantaged	●										●	●	
Understanding the relationship between cost of travel and demand for travel								●				●	
Enhanced user experience													
Understanding current user experience		●	●	●				●	●	●	●	●	●
Developing strategies and techniques for optimizing user experience		●				●			●	●	●	●	
Delivering the highest value for money for all transportation customers	●		●			●			●	●	●	●	●
Identifying customer perception of current transportation service		●					●		●	●	●	●	●
Assessing the effect of transportation as a service on the overall transportation system						●			●			●	●
Ensuring that travelers make the best use of the transportation system taking account of transportation demand and prevailing conditions		●		●		●	●		●	●	●	●	●
Understanding the relationship between high-quality traveler information and travel behavior		●		●		●	●		●	●	●	●	

Figure 5.6   (continued)

a series of implementable steps that can be measured very accurately, it might make it difficult to communicate the value and benefits of the investments. While projects and programs can be budgeted, managed, and implemented, it is the services that are enabled by such projects and programs that deliver value in terms of safety, efficiency, and enhanced user experience. While projects and programs must ultimately be defined to act as the manageable building blocks, there is much value in adopting a service evolution approach to the roadmap. Such an approach is illustrated in Figure 5.7.

Define Policy Objectives

An important starting point in the development of any investment program for smart cities revolves around the definition and agreement of the transportation policy and detailed objectives to be addressed by the implementation. The exact mechanism for the definition of and agreement on these objectives may vary from one city to another but is typically part of the transportation improvement plan (TIP) and long-range transportation plan (LRTP) development processes. Depending on the point in the cycle where the smart city investment program is being defined, it may be necessary to take a retrospective look at existing policy objectives or to look forward to the development of new policy objectives to be included in a new version of the regional transportation plans. Smart city objectives can be developed as a customized list using the list of objectives defined earlier as a starting point.

Note that each policy objective should have a measurable performance management parameter or analytic associated with it. The definition of both

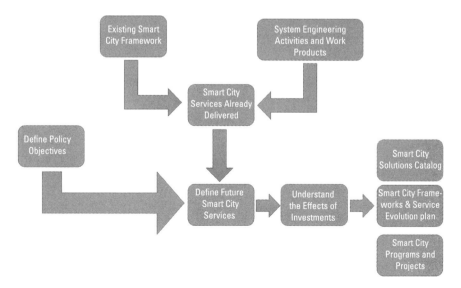

Figure 5.7   Smart service evolution approach.

the policy objectives and the analytics helps to support later efforts to determine the effectiveness of the smart city deployment using transportation analytics. At this point in the process, use cases will also be defined that explain the services to be provided and the data required and provide an estimate of the value and an indication of who will make use of the service. Building on these use cases, requirements can be defined to guide the development of the smart city framework. While defining policy objectives, it may also be the case that a coalition of partners required to deliver the smart city is established and managed—if a suitable coalition does not already exist.

Understand Possibilities

This will involve the development of a solutions catalog for the smart city. While the policy objectives and the use cases define what is required, the solution catalog defines how it can be achieved. In the development of the solutions catalog, partnership exploration can also take place to identify suitable private-sector partnerships that can be established to assist in the smart city delivery. Peer-to-peer information exchange with other smart cities may also be extremely valuable to understand what works and what does not work well developing the solutions catalog. At this stage, it would also be very useful to develop a funding source catalog that lists potential funding sources at the federal, state, and local levels as well as potential private-sector matching funds and initiatives.

Define Legacy

Prior investment in advanced technologies associated with transportation within the city must be considered when developing a smart city framework. In this step, existing investments are evaluated and translated into a range of smart city services that have already been delivered. Note that it is entirely possible that a partial implementation of the service has been implemented. In this case, service quality and coverage most also be evaluated. Advanced transportation technologies have been under development and deployment around the globe for quite some time.

However, mainstream application of advanced technologies to transportation really started in the 1980s. Therefore, it is highly likely that there will be some existing deployments within the city. It is necessary to define legacy systems and the results of prior investments to understand what has already been achieved and what can be used as a platform for future development. In many cases, multiple agencies may have been involved in the deployment of advanced technologies in the city, and the development of a comprehensive catalog will require information to be collected from each agency to develop a comprehensive catalog. The legacy catalog should address technology, organization, and business plan frameworks that are already in place. Once a catalog of deployments has been assembled it will be necessary to define the services that are being delivered by each of the deployments.

Define the Big Picture

Careful planning and preparation for the implementation of a smart city is particularly important. The type of technologies used include information and telecommunication technologies that lend themselves to being managed within a system engineering approach. In developing a plan for the future smart city, input should be received from a formal system engineering process that develops a system engineering management plan and a concept of operations for the smart city. These will help to define the technology framework of the big picture.

At this point organizational and business model frameworks will also be prepared that illuminate the shape of the future smart city framework from these perspectives. These frameworks will join the existing smart city framework and smart city services that have already been delivered to define future smart city services required. Taking input from the definition of policy objectives, the existing smart city framework, system engineering activities, and a catalog of smart city services already delivered, it is possible to define a range of future smart city services that are required to completely address the policy objectives, preserve legacy investment in technology, and comply with best practices in system engineering.

Define Implementation Plan

The smart city implementation plan should consist of service evolution, a legacy preservation and phasing plan, a financial plan, a marketing plan, and a risk management approach. The plan will also identify phase 1 implementations in terms of the services to be delivered and the programs and projects required to enable the services. In this step in the service evolution approach the value of big data and analytics comes to the fore.

Having the data available and analytics defined to understand the effects of investments provides a solid basis for the development of a smart city framework and as guidance for the service evolution plan. The proposed service evolution defined within the implementation plan will take full account of the effects of prior investments and make use of advanced data analytics techniques to gain insight into the likely effects of implementations and service deliveries.

This involves the definition of how a service will be implemented on an incremental basis within the region. The evolution of the service can be described in terms of two dimensions, time and space.

The term space is used as a generic term for simplicity and would in practice be replaced by a term that relates to the specific service. For example, for the traveler information service, the space parameter would, in fact, be the number of people that have access to the information services. The term space is used as a proxy for any parameter that describes the extent to which the service is available in the region. The term time is of course related to the period over which the service is evolved over the city. Once the evolution of services over a city has been defined, it is possible to define projects and programs that will result in the implementation of the service. The advantage of using the service evolution approach is ensuring that the implementation and funding are closely related to values and benefits. While projects and programs are required to implement services, it is the services that deliver the value and benefits. This service approach also facilitates marketing and public outreach to explain what is being done in the city what the value will be to the user.

Implement Phase 1

The implementation plan will have defined the first phase of implementation for the smart city. During phase 1 implementations it is important to attain early results, define the value being delivered, communicate success to partners and to the public, and monitor results. Formative and summative evaluation should be conducted on the phase 1 evaluation with formative evaluation used for project management and near real-time feedback, and summative evaluation used as a feed-forward mechanism to the next phase of deployment.

Learn

The results of the formative and summative evaluations are used to develop a series of practical lessons learned and experiences gained that will be used to inform subsequent deployment. This information will also be used to support outreach activities and to act as a guide for revising the big picture. This will result in the definition of a phase 2 implementation.

Continuous Smart City Implementation

Smart City implementation will continue in phases beyond phases one and two. The number of phases required relate to budget constraints and the availability of city resources to manage and deliver the projects.

5.7   Smart City Frameworks

In defining the big picture as part of the service evolution approach it is necessary to address technology, organization, and a business plan. It is worthwhile to consider this in more detail as these frameworks are crucial to the success of advanced technology implementations in past projects.

Figure 5.8 shows the technology framework developed by the City Protocol organization, which is based on that developed by the City of Barcelona [12]. The following sections examine each element of the architecture.

Smart Information Infrastructure

This is a collective term for the combined sources of data that will be provided to the event repository and big data management system. This could be raw data from sensors or probe vehicles or could be processed data, information coming from external systems

City Open Apps and Open Data

One of the important values of a centralized data repository or data lake is the ability to make access to the data available to a wide range of partners both in the public and private sector. Within the public sector, access to the data lake would enable each agency and each public-sector partner to develop its own analytics and strategies. With respect to the private sector, smart app developers and other private sector entities may use the data as the basis for new business opportunities and the development of consumer-oriented smart phone applications.

Citizens Participation Platform

The ability to channel data effectively and to turn into information can also be used with smart phone and web technology to support citizen participation. This is a two-way dialogue in which the citizen provides data and is in turn provided with information and the ability to comment on the quality of services.

Sensor Platform

This element supports the fusion of data from multiple sensors including infrastructure-based and probe vehicle–based sensors. It is also possible that the sensor platform will conduct a certain amount of preprocessing, often referred to as edge processing to optimize the use of communication networks as the data is transmitted to the back office or the city operating system.

Other Sensor Platforms

The smart city may also take advantage of other sensors that have already been installed by public and private sector entities. This could include sensors used for supervisory control and acquisition of data (SCADA).

Video Platform

This can be considered as a special type of data fusion, in which video images from a variety of sources can be brought together and managed. The video platform might also provide the ability to share data across partners.

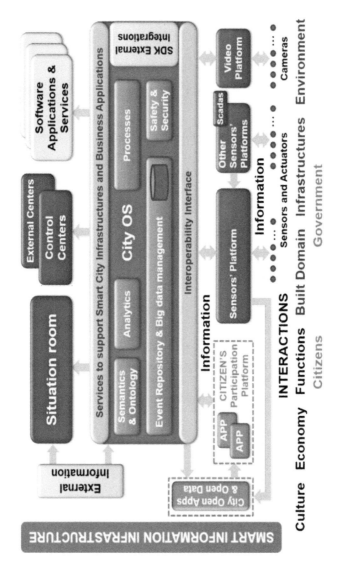

Figure 5.8 Smart city service technology framework.

Interoperability Interface

An essential ingredient in the smart city back office, the interoperability interface provides a means through which various data sources are ingested into the back office. The interoperability interface will also support the flow of data from the back office to a range of public- and private-sector applications.

Event Repository and Big Data Management

The smart city data lake is comprised of an extensive data repository that covers events and contains the data that describes transportation demand, prevailing transportation conditions, and supply characteristics for the smart city.

Safety and Security

Safety and security involves the application of dynamic safety and security functions that protect the confidentiality and integrity of data stored in the data lake and within the back office. Safety and security will also involve monitoring the interfaces that the back office uses to ingest and share data.

Software Development Kit (SDK) External Integrators

The software development kit provides an interface to external integrators such a smart application developers and other users of back-office data. This contains details on how to access the data, the format and content of the data, and other characteristics required to integrate data into applications.

Semantics and Ontology

Sometimes referred to as metadata, master data management, and data governance, this element provides the governance of data required to make sure that it is readily available to all users and that it is stored in an optimized structure within the back office.

Analytics

This describes the application of transportation analytics and other smart city analytics to gain insight and understanding through discovery and standardize business processes for the smart city. These can be used for discovery purposes and for business process management.

Processes

In addition to discovery analytics, a range of standardized processes are identified and supported. One of the essential ingredients for the successful application of technology to organizations is the identification of business processes and the customized application of technology to either support existing processes or implement new optimized processes.

Services to Support Smart City Infrastructure and Business Applications

A range of data services emanate from the smart city data back office to support smart city infrastructure, services, and business applications. It is likely that the data services will include both raw data and processed data, with the back office, in many cases, delivering added value.

Situation Room

This is the smart city nerve center that supports communication to all other modes' specific control centers and act as a nexus for developing response strategies.

Control Centers

Many mode-specific control centers such as traffic management centers, transit management centers, and event management centers already exist within the existing city infrastructure and can be incorporated into an overall framework.

Software Applications and Services

These are software developments that are customized to address specific applications and to deliver specific services. An example would be a connected citizen application that provides the smart city citizen (or smart citizen) with information regarding services that are available and prevailing conditions for transportation within the city. Another example would be a crowdsourcing application that enables smart city citizens and visitors to provide data to the back office and to provide data regarding their current perception of services within the smart city.

Note that this is an abstract architecture designed to illustrate and communicate the various elements that would comprise the technology framework for a smart city. It is not designed to be a practical implementable framework as considerably more detail would be required, including a concept of operations and a system engineering management plan, at the minimum.

As a complement to this technology view, the City Protocol team also developed an organizational view of a smart city [12], as shown in Figure 5.9.

This organizational view provides a framework or a checklist for defining roles and responsibilities with respect to governance and accountability; information and communication technologies; and the structure, interactions, and societal impacts of smart cities. This organizational framework, being fully compatible with the chosen technological and business model frameworks, would be a useful starting point for a smart city working toward customizing its own organizational framework.

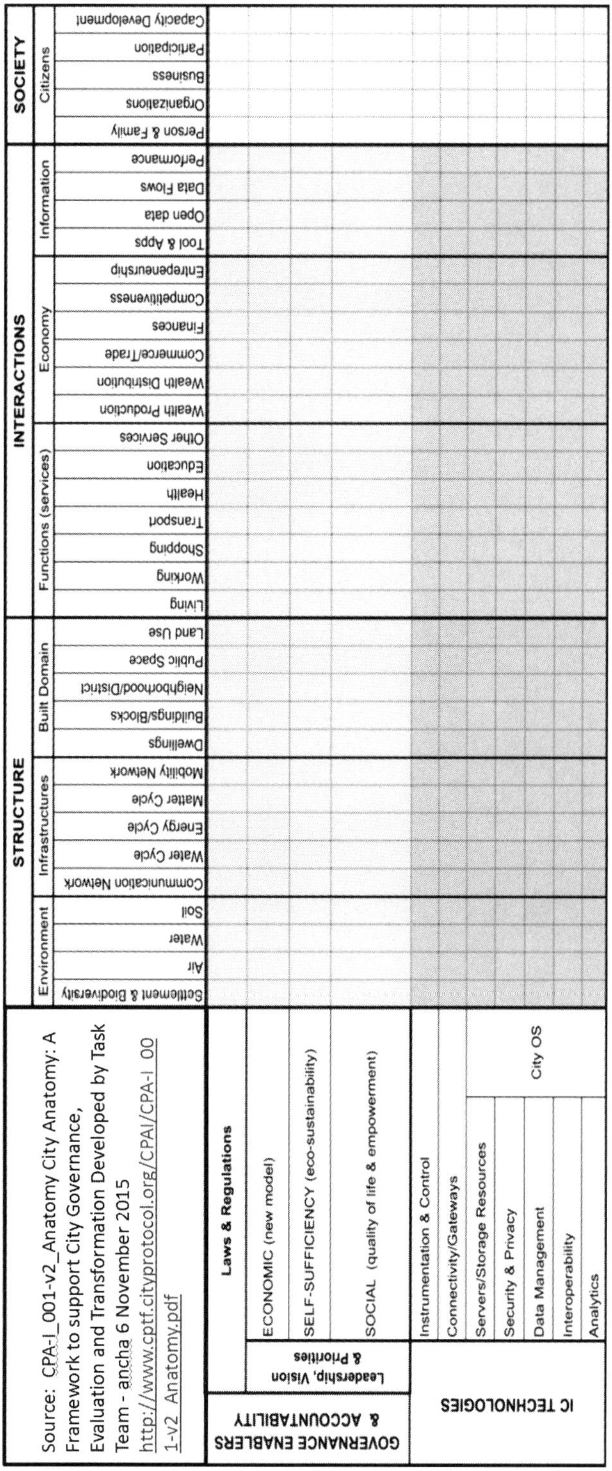

Source: CPA-I_001-v2_Anatomy City Anatomy: A Framework to support City Governance, Evaluation and Transformation Developed by Task Team - ancha 6 November 2015 http://www.cptf.cityprotocol.org/CPAI/CPA-I_001-v2_Anatomy.pdf

STRUCTURE

Environment: Settlement & Biodiversity, Air, Water, Soil

Infrastructures: Communication Network, Water Cycle, Energy Cycle, Matter Cycle, Mobility Network

Built Domain: Dwellings, Buildings/Blocks, Neighborhood/District, Public Space, Land Use

INTERACTIONS

Functions (services): Living, Working, Shopping, Transport, Health, Education, Other Services

Economy: Wealth Production, Wealth Distribution, Commerce/Trade, Finances, Competitiveness, Entrepreneurship

Information: Tool & Apps, Open data, Data Flows, Performance

SOCIETY

Citizens: Person & Family, Organizations, Business, Participation, Capacity Development

GOVERNANCE ENABLERS & ACCOUNTABILITY

Leadership, Vision & Priorities

Laws & Regulations:
- ECONOMIC (new model)
- SELF-SUFFICIENCY (eco-sustainability)
- SOCIAL (quality of life & empowerment)

IC TECHNOLOGIES

- Instrumentation & Control
- Connectivity/Gateways
- Servers/Storage Resources
- Security & Privacy
- Data Management
- Interoperability
- Analytics

City OS

Figure 5.9 Smart city organizationlal framework.

5.8   Evaluating the Effects of Investments

Evaluating the effects of investments in smart cities is an important element of planning and implementation. The ability to conduct before and after studies to understand the effects of implementations is supported by the availability of a coherent and comprehensive data source. One of the features of the application of intelligent transportation systems within North America and Europe has been the lack of a consistent framework for evaluation. To counter this deficit, smart city planning should incorporate a detailed definition of arrangements for performance evaluation, including the effects of investments.

The definition of the analytics to be used to measure the effects of investments is also of importance as this enables an assessment of the data required. In many cases, transportation data is collected on an ad hoc basis with levels of detail and quality that are unrelated to the intended purpose of the data. The development of a data collection plan that takes full account of the proposed use of the data is essential to ensure value for money and to enable a coherent assessment of the effects of the investment. A central data repository or data lake would be an essential ingredient to this approach. A data lake is constructed from multiple data sources and features the ability to support multigenre analytics, a function that addresses the challenge of separating the effects of different investments.

5.9   Smart City Challenges

There is a range of challenges and opportunities associated with the smart city. These are discussed in this section along with a summary of practical lessons learned from the London Congestion Charge project [13]. Note that this is not intended to be an exhaustive set of challenges, opportunities, or lessons learned, but can serve as the basis for the development of a checklist for a specific smart city implementation.

New Partners from Multiple Disciplines

The application of intelligent transportation systems has already shown that need to manage multidisciplinary groups including transportation professionals, automobile manufacturers, electrical engineers, system engineers, and a range of other disciplines that go beyond the conventional asphalt, concrete, and steel transportation projects. Within a smart city environment, the number of participants and the characteristics of these participants cover an even wider spectrum.

Participants can include staff from the mayor's office, economic development professionals, technology incubation leaders, renewable energy professionals, and information technology companies. Since a smart city encompasses services beyond transportation, the participants involved in the project will come from a more diverse background than those working on the typical application of advanced technologies to transportation. This professional diversity will require a special emphasis on clarity of communications with respect to the planned technology deployments and the definition of roles and responsibilities among the partners.

Avoiding the Development of a Set of Stovepipe Projects That Do Not Connect Successfully

To optimize the effect of a smart city implementation, it is necessary to ensure that all projects and programs fit within a coherent framework that ensures coordination and avoids conflict or duplication of effort. The need to achieve early results and shorter-term return on investment can often influence smart city initiatives toward high-impact investments in specific areas that lack coordination. The development of a big picture approach to a smart city plan and the definition of technology, organizational, and business plan frameworks are central to ensuring that the city is getting the very best from each investment and that each project builds on the other one.

Leveraging Sunk Investment in Legacy Systems to the Fullest Effect

While the destination point for a smart city journey will probably be very similar for all cities, the departure point can vary substantially depending on the perception of specific needs and prior investments in advanced technology. If an effective business justification is to be created from early investments in a smart city, it is important that prior investment in legacy systems are fully incorporated as a platform for moving forward. This requires an assessment and understanding of the services being delivered by legacy systems and the incorporation of relevant aspects into the technology and organizational frameworks.

Keeping the Focus on Results

A strong emphasis should be placed on defining and agreeing on objectives for the smart city initiative, the development of use cases and requirements, and the definition of arrangements for monitoring and managing the success of the deployment. It is likely that a smart city initiative, due to the variety of services to be offered, will have a wide range of objectives. Accordingly, it will be necessary to ensure that all planned actions and investments are related to these objectives and that all objectives can be addressed fully.

Communicating the Value and Benefits to Partners and the Public

Explaining the value and benefits of the proposed smart city elements to partners and to the public is one of the most important factors in the success of the initiative. Accordingly, the value proposition must be clearly defined and communicated through the appropriate channels to partners and public alike. The evaluation of benefits can be achieved through both summative and formative evaluation techniques. Summative evaluation techniques, as the name suggests, provide a retrospective summary of the benefits achieved from prior actions. Formative evaluation provides near obtained feedback that can be used to guide the initiative on an incremental basis.

5.10   Smart City Opportunities

This section provides examples of the opportunities associated with smart cities.

Smart City

Through the adoption of a coherent and coordinated approach to the delivery of smart city services it is possible to identify and take advantage of possibilities for synergy between services. For example, data collection services can provide data that is fed into a central data lake or repository that can support multiple applications. These applications could include connected citizens and connected visitor services. The data collection itself may be supported by crowdsourcing and the use of movement analytics. The creation and management of a data lake also ensures that a single version of the truth, with respect to data, can be created and maintained, enabling many services to take advantage of the data.

Cost Share on Project Implementation

Like most transportation implementations, one of the most important and pressing aspects of smart cities initiatives lies in the identification of funding. By developing a coordinated framework in a big picture for the smart city initiative, it is possible to identify opportunities to share cost on project implementations. Partners can identify opportunities to cost-share or to avoid investment costs by relying on services to be provided by another partner. This also enables the definition of partnership opportunities between the public and private sector.

The Ability to Ensure that Initiatives from Different Agencies Work Together and Do Not Conflict

Multiple partner agencies are likely to be involved in the deployment of a smart cities initiative. A high importance is placed on ensuring that the efforts of different agencies work together toward common goals. This avoids potential conflict between projects and maximizes the collective effect of individual projects. The adoption of a framework approach to Smart City planning supports a strong ability to optimize individual agency actions.

Improve Transportation Services for a Wide Range of Citizens and Visitors

Due to the broad nature of smart city services, there is an opportunity to improve transportation services for a wide range of citizens and visitors. Taking a comprehensive approach to the definition of objectives and proposed investments, it is possible to address a wide spectrum of transportation needs, including private vehicles, freight vehicles, transit vehicles, and transit for the transportation-disadvantaged. The smart city can also improve access to jobs, increase mobility in general, and take account of the need for social equity. Urban analytics approaches within a smart city can provide insight and understanding with respect to the factors that cause resistance to travel including trip time, trip time reliability, and model interchange times and can help assess the value proposition provided to all citizens and visitors.

Optimize Land Use

Land use tends to be a longer cycle time activity as investments in buildings and infrastructure typically have a design life of more than 50 years. However, a comprehensive approach to smart cities will include short-, medium-, and long-term initiatives and incorporate land-use changes and optimization into the overall smart city plan. Here again, urban analytics can provide new insight into the relationship between land use and transportation demand and support new insights and understanding based on observation.

Mitigate the Effects of Transportation on the Environment

Smart city transportation can affect the quality of life of citizens and visitors, through enhanced economic prosperity and the management of undesirable side effects. Transportation impacts on the environment include emissions, fuel consumption, noise intrusion, visual intrusion, and other factors that accompany the desirable effects of transportation. Through the application of advanced technologies and management solutions in a smart city, it is possible to deliver the desirable benefits while managing the undesirable side effects of transportation. This can include electric vehicles, optimized routing, and overall optimization of the multimodal smart city transportation system.

Gain Insight and Understanding into Transportation Supply and Demand and Prevailing Operating Conditions

Through the incorporation of smart sensors connected to a communication network, the use of probe vehicle data from connected and autonomous vehicles and the use of crowdsourcing techniques from smart phones, the smart city manager can have an unparalleled picture of prevailing operating conditions within the transportation network. This leads to an exceedingly rich data stream that can support the use of analytics for new insight and understanding into the quality of transportation service delivery, the factors that drive the demand for transportation, and changes in the ability to supply transportation services. A mix of short-, medium-, and long-term strategies can be developed based on this data stream to enhance transportation planning, to provide new services to the full spectrum of transportation users within a smart city, and to optimize transportation service delivery and operations.

Develop New Response Strategies

The possibilities to support the development of new response strategies within smart cities are immense. A scientific approach to transportation planning, traffic engineering, transit management, and mobility on demand services is enabled by new data and new data management techniques. While these new possibilities will initially impact planning and operations in terms of decision support to professionals engaged in these activities, it is also likely over the long term to lead to more automation. There is considerable current focus on the automated vehicle, and it is to be expected that the automated back office will also be developed as a complement to this.

5.11   Lessons Learned from the London Congestion Charge Project

In addition to the challenges and opportunities described above, there are several practical lessons that can be learned based on Transport for London's experience in implementing the London Congestion Charge, which was introduced in 2003 [13]. The congestion charge was introduced after a considerable amount of study on the issue of congestion charging or road pricing. The study started in 1964 with the publication of the Smeed Report [14], which presented the results from a government panel that assessed the practical issues related to the implementation of road pricing within a British city. Subsequent studies focused specifically on London, leading to the implementation of the congestion charge as transportation policy in London moved away from road construction in favor of public transport and traffic management improvements. Figure 5.10 [14] illustrates the current boundaries of the congestion charging zone. This represents one of the largest congestion charging zones in the world and was implemented with the general objectives of reducing traffic flows within the zone boundaries, while generating revenues for improvement to the public transportation system. Drivers are charged a fee of approximately $14 (all fees have been converted to U.S. dollars at the rate of $1.24 to the pound and rounded to the nearest dollar) for each day that their vehicle is detected as being present within the zone. The charge is in effect from 7 a.m.

Figure 5.10 Current boundary of the London Congestion Charging zone.

until 6 p.m. on weekdays. Charges must be paid by midnight on the day that the vehicle was detected inside the zone. The charge increases to approximately $17 if paid by the end of the following day. A discounted rate of $13 is offered if the driver chooses to pay with the automated payment system. Fees for late payment range from approximately $80 to $160, depending on lateness of the payment. The charge is enforced using automated number plate recognition technology. Video cameras are deployed at strategic locations within the zone, enabling Transport for London to detect vehicles approximately 11 times for an average trip through the zone. Gross revenue from the program during the first 10 years of operation was approximately £2.6 billion, of which 46% was invested in transportation improvements. A further 37% was also invested in improvements to the bus network.

While the charge was implemented more than 14 years ago and is a congestion-charging project rather than a full-fledged smart cities project, the lessons [15] are directly relevant to smart cities as they summarize practical challenges and opportunities associated with a real-world deployment of advanced technology in a major city. The lessons have been grouped into seven categories as illustrated in Figure 5.11, and the following sections describe lessons in each of these seven categories.

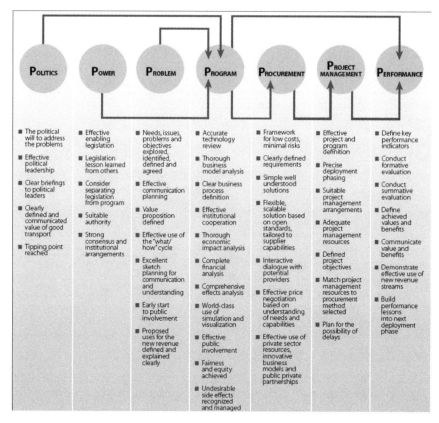

Figure 5.11   Lessons learned from the London Congestion Charge project.

Politics

This encompasses the ability to gain and retain political support for the project by communicating the value of the services to be delivered. A structured outreach program is required to initially communicate the value and benefits of the program to political decision-makers. The outreach must then continue to ensure that the political support gained during the initial communications is maintained throughout the life of the project implementation.

It should be noted that additional policy requirements can drive up the cost of the project. Transport for London's experience was that this type of project required a larger budget for communications and marketing; this was partly due to the need to explain the advanced technologies involved to the public and partly to explain why a congestion charge was necessary and how the revenue would be used to improve transportation. From a smart city perspective, it may also be necessary to allocate a larger budget for outreach and communications to explain the value and benefits and appropriate usage of the services to citizens and visitors.

Power

Transport for London attained a great deal of value by ensuring that legislation required to allow the project to be implemented had been addressed at an early stage. In London, the legislation required was defined, debated, and passed into law several years before the congestion-charging implementation was launched. The long study period before implementation afforded a public debate well ahead of the implementation decision, supporting a rational nonemotional discussion of the subject. From a smart city perspective, this could be directly relevant to regulations and legislation required for the use of autonomous vehicle technology within a smart city framework. Early discussion and establishment of such regulations not only supports a full discussion, but also provides the basis for public-private partnerships.

Problem

Clear concise definition of the problem in terms of the needs, issues, problems, and objectives to be addressed, was a feature of the London project. This included the definition of the value proposition for the end user. The development of a clear, structured, and agreed-upon problem statement also contributed significantly to the project's success. Accordingly, smart cities initiatives should follow suit, carefully defining objectives, use cases, and requirements.

Program

This includes the definition of a structured program to implement the project. A series of projects may be defined, phased, and linked together to achieve the overall objective. This requires a clear understanding of technology capabilities and the selection of a business model to ensure that technology capabilities are matched to the selected business model. This also requires effective communications and planning, strong institutional cooperation, a complete economic and financial analysis, and a detailed effects analysis. The development of technology, organizational, and business model frameworks for smart cities enables the coordination of various projects and ensures that all planned actions in investments can be linked back to objectives.

Procurement

The procurement approach should be selected based on minimizing life cycle costs and risks. It should incorporate clearly defined requirements and seek simple well understood solutions. As discussed earlier, requirements should be as free from ambiguity as possible. The objective of procurement should be to try to acquire products and services that are flexible and scalable and, preferably, that feature the use of open standards and architecture. An early interactive dialogue with potential product and service providers using a request for information process can be very enlightening. This provides information on current technology capabilities and limitations and helps to define service and product providers' capabilities and constraints. This can help to ensure that the requirements and the procurement documents are realistic and practical. Through the adoption of a big picture planning approach and an incremental implementation plan for a smart city, the most appropriate procurement mechanisms can be applied. Given the important role of the private sector, it may also be necessary to be innovative in the procurement process to support an effective dialogue regarding the technological possibilities and the best way to deliver services.

Project Management

Project and program definition includes the use of best practices for deployment planning and phasing. This ensures that sufficient project management resources are made available on both the public- and private-sector sides. This also requires a clear definition of project objectives and a contingency plan for possible delays. Many public agencies don't have the in-house expertise to plan, procure, and manage the deployment and implementation of a large Intelligent Transportation System. In many cases, specialized assistance is required. Several of the U.S. DOT Smart City Challenge applications feature the establishment of a smart city project management office to focus project management expertise and support coordination across all projects.

Performance

Performance management for smart city projects supports both summative and formative approaches. The summative approach would incorporate lessons learned at the end of the project and provide information to guide subsequent projects. The formative approach would provide information that can be used to guide the project in real time and to keep the project on course. In the transportation profession, performance is a term often used to gauge the efficiency of system operation. It is a metric used to verify that the system does what it is supposed to do. Performance metrics are also established to monitor and manage operations. It is also worth noting the difference between performance measurement and performance management. Performance management involves measurement, analysis, and the development of response strategies. An old management adage states, "If you are not measuring it you are not managing it." With respect to smart cities it is also appropriate to add, "…and if you are only measuring it you are still not managing it." Big data in the form of an appropriate data lake and the use of transportation data analytics can significantly improve performance management.

5.12 The Sentient City

Section 5.4 includes a rather light-hearted definition of a smart city, which I'll repeat here: "You know your city is smart if you poke it with a stick and it reacts appropriately." This definition makes the point that a smart city should have the ability to sense and the intelligence to react appropriately based on the results of the sensing. This requires a sophisticated ability to digest a wide range of data and to identify opportunities and threats and respond appropriately. It also requires the successful incorporation of data and technology into governance and business processes.

The concept of a sentient enterprise was first defined by Mohan Sawhney (McCormick Foundation chair of technology and clinical professor of marketing and director of the Center for Research in Technology & Innovation, Kellogg School of Management, Northwestern University) and Oliver Ratzesberger, Teradata executive vice president and chief product officer [16]. Figure 5.12 illustrates how the sentient enterprise defines an organization as a single organism that is capable of feeling, perception, and self-awareness. This envisions a future enterprise that can listen to data in real time and develop responses based on intelligence.

A discussion regarding these abilities is currently under way within the private sector as there is a growing understanding of how big data and analytics can be harnessed to provide the level of sensing and intelligence required to support enterprise agility and robust operations. The sentient enterprise concept could also point the way toward the future of smart cities.

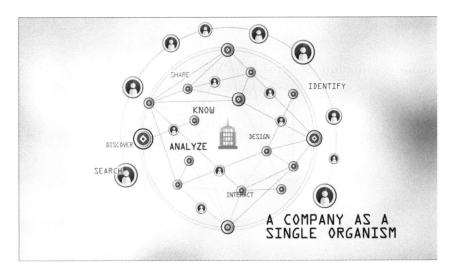

Figure 5.12 A company as a single organism—the Sentient company.

The increasing sophistication and capability of data science provides some fascinating possibilities for the future smart city to incorporate sophisticated sensing and migrate toward the automation of many functions. While the sentient enterprise will be focused on monetizing results, the sentient city will be focused on optimizing safety and efficiency and enhancing user experience.

5.13 Summary

This chapter provides an overview and a working definition of a smart city. It defines a range of objectives that can be addressed by a smart city from a transportation perspective, with examples of the steps that can be taken toward implementing a smart city in a structured and risk-managed manner. It also discusses the concept of a framework approach to planning to support project and investment coordination and offers approaches for coordinating smart city investments. In addition, it describes approaches to the evaluation of smart city investments and the effects of initiatives.

Based on practical experience from current smart city initiatives and from previous investments in the application of advanced technology to transportation, the chapter identifies a range of challenges and opportunities and presents some practical lessons learned.

The chapter concludes with a look to the future on how the current smart city might evolve into a sentient city that has the capability of sophisticated sensing and the intelligence to develop appropriate responses in a more automated manner.

It is obvious that the smart city concept has focused our attention on the application of technologies within organizational and business model frameworks. We are learning that success requires the effective application of technology in a manner that takes account of the human dimensions.

References

[1]   Google Trends, https://www.google.com/trends/explore?date=today%205-y,today%20 5-y&geo=,&q=Smart%20cities,intelligent%20transportation%20systems, retrieved September 18, 2016.

[2]   Report to the President, Technology in the Future of Cities, Executive Office of the President, President's Council of Advisors on Science and Technology, February 2016, https:// www.whitehouse.gov/sites/whitehouse.gov/files/images/Blog/PCAST%20Cities%20Report%20_%20FINAL.pdf, retrieved September 28, 2016.

[3]   U.S. DOT, Smart City Challenge Phase 2: Notice of Funding Opportunity, 18 April 2016, https://www.transportation.gov/smartcity/nofo-phase-2, retrieved September 28, 2016.

[4]   Smart Cities Readiness Guide, Smart Cities Council, 29 October 2015, http://smartcities-council.com/resources/smart-cities-readiness-guide, retrieved September 28, 2016.

[5]   Austin Vision Narrative—Vision for a 21st-Century Mobility System, https://www.trans-portation.gov/smartcity/visionstatements/index, retrieved September 28, 2016.

[6]   Columbus Ohio Vision Narrative—Columbus Smart City Application, https://www.transportation.gov/smartcity/visionstatements/index, retrieved September 28, 2016.

[7]   Denver Vision Narrative—Beyond Traffic: Denver of the Smart City Challenge, https://www.transportation.gov/smartcity/visionstatements/index, retrieved September 28, 2016.

[8]   Kansas City Vision Narrative—Beyond Traffic: The Vision for the Kansas City Smart City Challenge, https://www.transportation.gov/smartcity/visionstatements/index, retrieved September 28, 2016.

[9]   Pittsburgh Vision Narrative—City of Pittsburgh Proposal beyond Traffic: The Smart City Challenge, https://www.transportation.gov/smartcity/visionstatements/index, retrieved September 28, 2016.

[10]   Portland Vision Narrative—Ubiquitous Mobility for Portland, https://www.transportation.gov/smartcity/visionstatements/index, retrieved September, 2016.

[11]   San Francisco Vision Narrative—City of San Francisco Meeting the Smart City Challenge, https://www.transportation.gov/smartcity/visionstatements/index, retrieved September 28, 2016.

[12]   CPA-I_001-v2_Anatomy City Anatomy: A Framework to support City Governance, Evaluation and Transformation Developed by Task Team—ancha 6 November 2015 http://www.cptf.cityprotocol.org/CPAI/CPA-I_001-v2_Anatomy.pdf.

[13]   Wikipedia, London congestion charge, https://en.wikipedia.org/wiki/London_congestion_charge#cite_note-41, retrieved April 8, 2017.

[14]   The boundary of the current congestion charging zone, CC BY-SA 2.0, https://commons.wikimedia.org/w/index.php?curid=1627312.

[15]   Interview with Jeremy Evans, then head of technology, Transport for London, August 10, 2006.

[16]   Ratzesberger, O., *Teradata Presentation,* "Overview of the Sentient Enterprise," https://teradatanet0.teradata.com/docs/DOC-85836, retrieved September 28, 2016.

6

What Are Analytics?

6.1 Informational Objectives

This chapter answers the following questions:

- What are analytics?
- Why are analytics valuable?
- What's the difference between analytics reporting and key performance indicators (KPIs)?
- What are some examples of analytics?
- What are some examples of analytics applied to transportation?
- What's the best way to use analytics?
- How do analytics and data lakes fit together?
- How do we identify data needs associated with analytics?

6.2 Introduction

The application of analytics to transportation within a smart city is a new subject. Consequently, reference material and literature available on the subject is scarce. This chapter addresses the need for more information by defining and describing analytics from a smart city transportation perspective. The effective use of analytics has the potential to revolutionize transportation, by providing deeper insights and greater understanding than we have ever been able to achieve. The converging availability of big data, suitable analytic techniques, and data can unleash the power of big data. This has the potential to provide new insights and understanding regarding transportation supply and demand. It is likely that the impact of analytics on transportation will be as great as that of the Internet. In the same manner as we approach web services, however, care must be taken if the best results are to be achieved.

This chapter discusses the nature of analytics by providing an overview of analytics and an exploration of how analytics can be applied across the range of smart city transportation services defined in Chapter 5. The chapter also discusses analytical performance management for smart cities, explaining the relative merits of KPIs and analytics. The chapter concludes with a discussion of the relationship between analytics and data. This illustrates the power that can be unlocked through the application of a combination of big data and analytics.

6.3   Chapter Word Cloud

The word cloud in Figure 6.1 provides an overview of the content of the chapter by listing the most frequently used words, with the font size proportional to the frequency of use of the word. As the word cloud provides some insight into the characteristics of the chapter at a glance, it could be considered to be an analytic. Section 6.6 discusses the use of analytics to characterize smart city transportation services.

Figure 6.1   Word cloud for Chapter 6.

6.4 What Is an Analytic?

Big data in itself contains little value. The value is there, but it is latent and must be activated. Some hold a low opinion of the value of data for this reason. It is important, however, to think of data as a raw material, from which something very valuable can be created. The real value in collecting and managing big data lies in the conduct of analytics that turn the data into information, insights, and actionable strategies. This chapter explores the definition of data analytics. In particular, it explains the difference between world-class reporting and the use of analytics. Making use of a sporting analogy, world-class reporting will only ever make you a well-informed spectator at a football match, whereas analytics can give you the power to change the performance of the team, just like a coach. Examples of analytics that have been applied to transportation and to business enterprises beyond transportation will be explained later in Section 6.6.

A more formal definition of the term analytic can be found in the *Oxford English Dictionary* [1], in which analytic has the following meanings:

- "The branch of logic which deals with analysis (see analytics n. 1a) (obs.); an analytical system, method, or approach; an analysis."
- "Of, relating to, or in accordance with analysis or analytics; consisting in, or distinguished by, the resolution of compounds into their elements."
- "Of a judgment, statement, proposition, etc.: expressing no more in the predicate than is contained in the concept of the subject; true simply in virtue of its meaning or its logical form; having the property that its denial is self-contradictory."
- "Forming part of mathematical analysis (analysis n. 5); relating to or involving mathematical analysis."
- "Of a function: having derivatives of all orders at every point of its domain (or a specified part of its domain); locally representable by a power series."
- "That analyses or has a tendency to analyse; that is concerned with or characterized by the use of analysis."
- "Characterized by the use of separate words (auxiliaries, prepositions, etc.) rather than inflections to express syntactical relationship."

These definitions indicate that the word analytic can mean a system and method or approach to analysis, fundamental truths, or a mathematical analysis and breaking down of something into its individual elements. From a data perspective, an analytic provides insight and understanding that can be used to improve the performance of a transportation enterprise, process, or delivery of service.

Typically, the word analytic is used interchangeably with performance management parameters and KPIs. The two are typically used in association with performance reporting and measurement. The difference between these terms and the term analytic lies in the ability that an analytic provides to go beyond reporting and measurement, allowing for insight that can form the basis for actionable strategies that will affect performance, rather than just measuring it.

Promoting smart city managers and transportation professionals from spectators to coaches can be achieved through the effective use of analytics. This is the added value that is sought from analytics and that drives the desire to harness the power of big data and analytics in transportation today. There are many new technologies and approaches available in transportation. Combined with the more traditional asphalt, concrete, and steel projects, they create value and benefits. In order to guide the application of constrained resources into these areas, it is necessary to have as full an understanding as possible of the effects of prior and future investments. The insight and understanding that can be obtained from the appropriate use of analytics will have a significant impact on planning, design, delivery, operations, and maintenance of all aspects of transportation within a smart city. The nature and characteristics of analytics are discussed in this chapter, within the context of a smart city.

6.5   Why Analytics Are Valuable

The value of analytics lies in the ability to glean new understandings, to identify trends, and to reveal patterns from a big data set. The larger and more varied the data in the data set, then the higher the probability of new understandings and insights. Analytics also deliver another value in the form of the ability to extract information from and unlock the power of big data.

As the application of analytics to transportation progresses, it can be expected that *wow* and *whoops* moments will be encountered. A *wow* moment is when the use of analytics reveals a new data relationship, trend, or pattern that we did not know about previously. For example, we may find new relationships between the level of tolling on toll roads and the volume of traffic on the road. At the current time, a simple lookup table that relates toll level to expected traffic volume is used. In the future we may also take account of trip purpose, weather, and the driver's perception of alternative routes and modes.

A *whoops* moment may be encountered when new insight and understanding reveals deficiencies in the planning or delivery of smart city transportation services. This does not need to be the end of the world, but rather the beginning of a suitable response. A suitable response to new insight and understanding with respect to service issues and problems would be to develop a plan for addressing the issue and putting together a related budget. This is simply an evolution into scientific investment planning based on the identification of issues and an understanding of the effects of the investment.

Vast quantities of data present the challenge of making sense of the data, converting it into information, and doing something useful with it. The real value of analytics lies in the end result. As illustrated in Figure 6.2, there is a process involving analytics that should result in the delivery of practical value. The vital role of analytics lies in the conversion of the data to information, but even information cannot deliver value unless it supports actions and changes because of the new information.

Early in the development of intelligent transportation systems, many transportation professionals were concerned that new sensor and telecommunication abilities would reveal issues that could not be addressed due to budget constraints. This is another reason why the application of analytics must support the entire value chain from data collection, through information processing to the definition and application of response strategies. A comprehensive approach to the application of analytics across the full spectrum of transportation activities from planning through design delivery and operations will also help to focus available resources accurately and consistently. This should realign budgets and release funds for the issues uncovered by new insights.

Figure 6.2 is drawn as a pyramid to indicate the dependence of the success of later actions on the success and completeness of the earlier ones. Good data leads to good information and with the help of analytics, information yields actionable insights. It is also important to consider all the steps in the process in order to achieve results.

In practice, analytics will show the relationship between different pieces of data in the data lake, and it is this connection that can yield new results and

Figure 6.2 Analytics value chain.

understanding. For example, a retailer can gain insight into customer behavior through analytics that show that when customers buy one product they also buy another one. The action that could be taken as a result of this insight would be for the retailer to place both products adjacent to each other within the store to make it easier for customers. The retailer might also decide to place another product adjacent to the other two in order to increase sales of the third product. Analytics can reveal that the customer might have a need for the third product. Understanding the connection between the data items based on analytics allows new value to be realized.

Major banks also make use of analytics to determine the reasons why customers move to other banks. The path that a specific customer takes through customer-service facilities online and by telephone might reveal causes that can be rectified.

In another example beyond transportation, a well-known movie studio made use of analytics conducted on a sample of the global Twitter feed to assess the impact of a new movie, in terms of audience reaction. The insight gained was used as input to the decision on the timing for releasing the movie in theaters and subsequently moving the distribution of the movie to Netflix and cable.

The value of analytics also lies in the ability to go beyond reporting. Even world-class reporting, as discussed earlier, can only inform and provide summaries of measurements. Analytics can characterize trends, patterns, and insights within the data and provide the basis for actionable strategies that can improve the performance of the organization. Like the objectives and use cases discussed in Chapters 2 and 9, analytics have real value when applied within an overall process that leads to changes in the performance of the enterprise. This makes it very important to make use of analytics within an overall structured approach to extracting value from big data.

6.6   Smart City Services Analytics

Probably the best way to explain the application of analytics to smart city transportation is to provide some examples. Chapter 5 discusses the smart city from a transportation perspective and defines services that represent smart city transportation. These sixteen services are summarized in Table 6.1 along with analytics that has been defined for each service. This linking of services and analytics is intended to reinforce the notion the analytics should be identified based on objectives that lead to the definition of uses cases. This ensures that the initial objectives and desired end results are kept to the forefront as the analytics are identified and applied. While this process is an important aspect of analysis, it is not the only way to approach big data and analytics these days. The emer-

Table 6.1

Proposed Analytics for Smart City Services

Services	Analytics
Asset and maintenance management	Asset performance index, asset maintenance standards compliance measure, optimal intervention point analytic
Connected vehicle	Lane changes per mile, steering angle compared to road geometry, brake applications per mile, driving turbulence index, minutes per trip, trip time reliability index, number of stops per trip
Connected, involved citizens	Citizens' awareness levels index, citizens' satisfaction levels
Integrated electronic payment	Transit revenue per passenger, transit seat utilization, toll revenue per vehicle and per trip, premium customer identification index, parking revenue per slot, payment system revenue achieved compared to forecast and addressable market
Intelligent sensor–based infrastructure	Data quality index, transportation conditions index, trip time variability index
Low-cost, efficient, secure, and resilient ICT	Network load compared to capacity index, network latency, cost of data transfer, network security index
Smart grid, roadway electrification, and electric vehicles	Electric vehicle charging points per mile, electric vehicle charging points per head of population, number of electric vehicles as a percentage of the total fleet, electric vehicle miles per day, electric vehicle miles per trip, electric vehicle miles between charges
Smart land use	Observed trip generation rates for different land uses, observed actual trips between zones, land value transportation index, zone accessibility index
Strategic business models and partnering	Percentage of private sector investment, number of partnerships, improvement in service delivery for each private sector dollar invested
Transportation governance	Transportation efficiency for each dollar spent, supply and demand matching index, transportation agency coordination index, partnership cost-saving index, cost of data storage and manipulation compared to services provided
Transportation management	Mobility index, citywide job accessibility index, citywide transportation efficiency index, reliability index, end-to-end time including modal interchanges index
Traveler information	Traveler satisfaction index, decision quality information index, behavior change index
Urban analytics	Number of analytics in use, value of services managed by analytics, money saved through efficiencies gained by analytics
Urban automation	Percentage of automated vehicles within the entire citywide fleet, percentage of automated vehicles in use by city agencies and private fleets, proportion of deliveries made by automated vehicles, proportion of passengers carried by automated transit
Urban delivery and logistics	Average cost of urban delivery, average time for end-to-end delivery, freight and logistics user satisfaction index, freight management satisfaction index
User-focused mobility	Citywide mobility index, user satisfaction index, transportation service delivery reliability index

gence of discovery tools that can be used on a big data set to uncover trends and patterns within the data provides us with a flexible approach to analytics. Predefinition of analytics is like having a hypothesis that will be proven by the data analysis. Another way to approach this is through a discovery process that enables the data to talk, revealing further insights. This type of approach might yield suggested analytics that could be used in addition to the ones that have been predefined. The process of discovery for transportation holds the promise of a fascinating role for a transportation data analyst. Using a big data set, combined with a powerful discovery tool, should make it possible to build a deeper understanding with respect to transportation within a city or region. This will require a skill set that combines an understanding of data science with an understanding of transportation, a blend that seems to be rare at the moment. One way to address this would be to ensure that suitable data analytics expertise is built into a smart city planning and deployment teams. While in the short term executive-level staff and managers would not be expected to have a deep understanding of data science, they need to be supported by someone that does. Over a period of time, data science knowledge and awareness should grow at the executive level as the tools become more intuitive, enabling executive-level and management staff to conduct their own discovery and analytics. Perhaps, in the early days at least, this is an ideal role for a consultant.

Again, Table 6.1 shows a series of analytics for each of the 16 services identified in Chapter 5. The focus is placed on services since the concept of smart cities is to take full advantage of technology to provide better services to city residents and visitors. It is intended to act as a starting point by providing a sample of analytics in order to illustrate the nature and characteristics of analytics. It is expected that a smart city team would build on these and develop a set of customized services and analytics specifically for the city in question. It is also likely that analytics for one service will be combined with analytics for another service to create hybrid analytics that will address either one service, or the other, or both. For example, trip time reliability from the intelligent sensor–based infrastructure service could also be used to measure the performance and effectiveness of other services. It is also likely that analytics derived for use in integrated payment, connected and autonomous vehicles, and travel information will find application in transportation performance management. This is another reason for taking a structured, coordinated approach to the definition of analytics for each service rather than an ad hoc one.

Asset and Maintenance Management

Analytic applied to asset and maintenance management associated with smart city transportation services will have the potential to improve the efficiency of service delivery. This will, in turn, improve the effectiveness and efficiency of the other services that are supported by asset and maintenance management.

"Mean time before failure" is a parameter often quoted with regard to operational performance of intelligent transportation system devices. This provides a measure of a single aspect of asset performance and a single data element. An analytics approach would produce a wider view of asset performance by taking into account multiple factors such as cost versus performance index, cost of maintenance, and overall design life in addition to the meantime before failure. This index would take account of partial as well as total failures. The index would be used to identify poorly performing devices and as input to an overall maintenance and replacement strategy; a smart city would also define and agree to set performance targets and maintenance standards for critical assets. Analytics can be identified to compare actual maintenance standards against those implemented and to generate an index that compares the total cost of maintenance for each device compared to the value delivered by the device. This approach can extend to all assets within a smart city including devices, telecommunications assets, vehicle assets, and all other assets associated with the delivery of the 16 transportation services identified for smart city implementation. The ability to identify trends and patterns with respect to asset performance and expenditure on asset maintenance will be a powerful element in the smart city due to the ability to ensure that resources are being utilized effectively and that value for money is being achieved. It is to be expected that life-cycle analytics will also reveal that the cheapest assets (low initial capital investment) may not present the best value for money over the life of the asset.

Connected Vehicle

It is anticipated that the connected vehicle will provide a richer stream of data in a big data set that features volume and velocity—with respect to data, that is, not the vehicle. This will allow the identification use of analytics that make comparisons between different data elements. For example, usage-based insurance professionals would be very interested to measure the number of lane changes made by a vehicle for every mile. This combined with steering angle compared the road geometry could provide some valuable insight into driver behavior and consequent risk. The number of brake applications and accelerator depressions per mile would form the basis for a driving turbulence index that could form the basis for crash prediction and certainly provide valuable information on the effectiveness of traffic signal timings and the number of minutes taken for each trip across the city; the reliability or variability of each trip would provide valuable input into overall transportation performance through the use of trip time and trip time reliability indexes.

Connected, Involved Citizens

The notion of a connected and involved citizen implies the ability to have a two-way dialogue with the citizen. In one direction, crowdsourcing, movement analytics, and social media data might be retrieved from the citizen. This will of course be summarized and anonymized before use. In the other direction, services and information can be pushed to the citizen to enhance smart city living and provide the equivalent of a user manual for smart city transportation services. This will be particularly important as the concept of MaaS is implemented, when citizens will be offered a range of mobility choices across multiple modes and from both the public and private sector.

This two-way dialogue will be enabled by the use of wireless and wire line communications networks that will link citizens to back offices and ultimately to the IoT, which connects a wide range of devices and appliances, along with connected vehicles, to a single network of networks.

The dialogue will support the use of citizen awareness analytics that can provide continual measurement regarding the perception of citizens with respect to services and the awareness of citizens with respect to available services. These can include trip modal choices and route and timing choices. A citizen satisfaction survey can also be conducted on a rolling basis making use of the two-way communications to determine how citizens feel about the quality of transportation service within the city and at given locations at any given time. This could be particularly useful for the evaluation of maintenance of traffic through major reconstruction zones, for example.

Integrated Electronic Payment

Citywide integrated electronic payment that supports payment of tolls, purchase of transit tickets, and payment of parking fees as well as payment for government services could support a number of financial analytics that measure the effectiveness and efficiency of the payment system. These could include the total volume of transit revenue per passenger, transit seat utilization, toll revenue per vehicle and per trip, and the identification of premium customers for all modes of travel. This latter would be achieved through a combination of electronic payment system and origin and destination data that would reveal the most valuable routes and customers.

Other analytics could be used to identify total citywide payment system revenue achieved compared to forecast and compared to the size of the addressable market. Due to the volume of data regarding prevailing transportation supply and demand, it is likely that this service will also support analytics for transportation performance management and travel information, at a minimum.

Intelligent Sensor–Based Infrastructure

Intelligent sensor–based infrastructure will allow multiple readings to be taken of the same data element and enable what is known as orthogonal sensing. This would support the definition of a data quality index that provides detail on the quality of the data being collected in terms of accuracy and completeness. A transportation conditions index could also be created from data that emanates from sensor-based infrastructure that shows the ebb and flow in the quantity and quality of transportation services being provided within the smart city. This could be supplemented by time and trip time variability indexes. Sensor data could also be combined with probe data from movement analytics and connected and autonomous vehicles to provide hybrid analytics.

Low-Cost, Efficient, Secure, and Resilient ICT

Assuming that low-cost, efficient, secure, and resilient ICT also includes management capabilities to measure the volume and use of each datalink, then analytics can be created to compare the total network capacity to the load on each link any given time. Network latency on a total network and individual link basis along with the cost of data transfer and a network security index could also be determined to measure the performance and gain insight and understanding into the delivery of information and communication technologies. This will support the application of network management techniques to transportation communication networks. It is also interesting to note that such techniques are likely to be applied to transportation services within the smart city in the future. The concept of a network manager is long established for computer and energy networks. The availability of data and analytics should make it possible to identify and support a "transportation network manager" role for the future smart city. It is now feasible, with the help of big data and analytics, to manage transportation service on a network and citywide basis.

Smart Grid, Roadway Electrification, and Electric Vehicles

The smart grid, roadway electrification, and electric vehicles service involves the use of electricity as an energy source for vehicles in a smart city. Analytics that can be used to characterize this service include those that relate to the availability of electric vehicle charging points and those that relate to the performance of the electric vehicles. For example, an analytic that defines the number of electric vehicle charging points per mile could be used to define the viability of electric vehicle operation in a smart city. Another analytics measure—electric vehicle charging points per head of population—could also be used to define the progress being made toward making electric vehicles ubiquitous in a smart city. The number of electric vehicles as a percentage of the total fleet, electric vehicle miles per day, electric vehicle miles per trip, and electric vehicle miles between charges could also be used as analytics to define the performance of the electric vehicle. An overarching analytic for the entire electric vehicle system would be the amount of energy being consumed for all vehicles over the entire city, compared to the degree of mobility made available by the service.

Smart Land Use

It has always been understood that there is an extremely close relationship between land use within various zones of the city and the demand for transportation. This is currently determined through the use of mathematical simulation modeling techniques. The use of urban analytics will enable us to make use of observed data regarding the demand for transportation and the behavior of transportation users in the smart city. Observed trip generation rates for different land uses can also be determined from national probe data and data from infrastructure-based sensors to enable us to have a high-resolution picture of the demand generated by each zone within a smart city. Observed actual trips between zones can also be combined with data regarding retail transactions to go beyond the definition of volume and to provide additional information on why travelers are traveling. These analytics could be combined into a single land value transportation index that would characterize the change in the value of the land within the zone in relation to the mobility and accessibility associated with the zone. This could also support the definition of a zone accessibility index for various trip purposes such as work, education, and leisure.

Strategic Business Models and Partnering

Strategic business models and partnering refers to an indirect set of services to be provided in the smart city. These could be viewed as enabling services that provide support for the services that deliver direct transportation. Business models and partnering can be characterized by an analytic that shows the proportion of private sector investment compared to total investment in a smart city. Another analytic that defines the number of partnerships compared to the improvement in service delivery for each private sector dollar invested could also shed light on the efficiency of the partnerships. This is an important aspect of smart cities that does not deliver smart city transportation services directly, but that can have a significant impact on the efficiency and effectiveness of service delivery. Public private partnerships for toll roads have also demonstrated that effective engagement of the private sector can substantially accelerate the deployment of projects and technologies. Such partnerships are likely to involve the work required to establish smart city transportation services and the resources required to deliver them.

Transportation Governance

The efficiency and effectiveness of transportation governance could be addressed through a number of different analytics. For example, an analytic that defines transportation efficiency for each dollar spent could be used to show the relationship between investment and improvements in transportation efficiency. Another way to measure the effectiveness of transportation governance would be to show how closely supply and demand are being matched. This could be represented by a supply and demand matching index that shows the difference between supply and demand for transportation in the smart city at any given time. Transportation governance should also include responsibility for coordination between the activities of various agencies that deliver transportation the city. This could be addressed by a transportation agency coordination index that defines the effectiveness of coordinated plans and activities between agencies. With respect to public-private partnerships a partnership cost-saving index could be used to show the financial advantage of public-private partnerships. With respect to big data, the cost of data storage and manipulation could be compared to services provided in an analytic that measures how effectively the entire city is collaborating on the use of big data. This is another crucial aspect of success in transportation delivery for a smart city as the technologies and services involved are best utilized in a sharing and coordination environment. Transportation has historically been operated by a number of highly focused agencies that derive great efficiency from specialization. Coordination and partnership must be added to this specialism if smart city transportation services are to be delivered in the most effective manner. This may require a reorganization and redefinition of arrangements for governance and management to accommodate the new challenges. Existing transportation agencies also require a certain degree of autonomy to implement policies and procedures without reference to other organizations. A balanced approach to autonomy and coordination will be required.

Transportation Management

It is expected that within a smart city, transportation management will be conducted on a multimodal basis across all transportation services delivered within the city. This will require the use of analytics that characterize the effectiveness and efficiency of all services. One such analytics could be a mobility index that measures the real mobility to and from each zone within the smart city in addition to an overall mobility measure for the entire city. A citywide job accessibility index analytic could also be used to characterize the ease or difficulty associated with trips to and from work. These analytics could be used in combination with a transportation efficiency index, a travel time reliability index, and a total travel time index to develop a complete picture of transportation management efficiency within a smart city. At the highest level analytics could be defined that compare the volume of public and private sector investment to the results obtained. This would include the application of scientific investment planning techniques discussed earlier in Section 6.5.

Efficient transportation management would also address parking management as part of the analytics approach. Analytics could be applied to revenue management, parking space utilization, user satisfaction with parking information, and parking supply planning topics, at a minimum.

Effective transportation management must address the same challenges as defined under transportation governance. Services are currently delivered by a range of autonomous transportation agencies, with the exception of a few of the largest cities in the world, where a single unitary transportation authority has been established. Examples include Seoul and London. As smart cities will face the additional challenge of integrating other services, such as smart energy and smart places to live and work, with transportation, there may be a need to revisit transportation management models within smart cities.

Traveler information services within a smart city will involve the delivery of decision-quality travel information to both citizens and visitors. A traveler satisfaction index could be used to measure citizens' and visitors' perceptions of the quality of traveler information services. A decision-quality information index could also be used to characterize the effectiveness of the travel information in terms of change and efficient use of the services within the smart city. This type of sophisticated decision-quality information delivery could be viewed as the equivalent of a user manual for the smart city transportation network, or as a soft form of transportation management that influences user behavior and makes system use more effective.

Urban Analytics

Urban analytics are used to characterize trends, patterns, and insights gained from the big data set collected by a smart city. However, this does not mean that we should not apply analytics to the performance of the analytics. Suitable analytics to characterize the performance of urban analytics would be the number of analytics in use, the value of services managed by analytics, and money saved through efficiencies gained by analytics. The total cost of applying urban analytics could also be compared to the value delivered and the money saved through the use of the analytics.

The definition of costs and benefits for smart cities, based on prior implementation experience, is a very important aspect of the business and financial aspects of the smart city. The identification of early winners along with the quantification and estimation of the value that can be realized is central to the justification of further investment in smart city analytics. This is discussed in more detail in Chapter 11.

Urban Automation

Urban automation includes the use of automated vehicles for transit, freight, and private vehicle travel. The main reasons to apply urban automation to the smart city would be to achieve benefits in terms of safety, efficiency, and enhanced user experience. Therefore, analytics could be defined for each category. With respect to safety, the number of crashes avoided or mitigated compared to the investment in urban automation would yield insight into improvements in safety. Defining another analytic that compares such reductions with investment in safety-related systems would also yield insight into the performance of such investments.

With respect to efficiency, an overall improvement in trip times and trip time reliability including wait times and modal transfer times would provide insight into efficiency improvements. Enhanced user experience could be characterized by the use of a user perception index supported by smart phone apps or the in-vehicle unit within the automated vehicle. Other analytics that would characterize the progress being made toward full automation of the city would be the percentage of automated vehicles within the entire citywide fleet, the percentage of automated vehicles in use by city agencies and private fleets, the proportion of deliveries made by automated vehicles, and the proportion of passengers carried by automated transit vehicles. These would all take account of the resources invested in the services and the availability of the services over time, space, and quality levels within the city.

Urban Delivery and Logistics

Analytics to characterize urban delivery and logistics would address cost, time, and reliability of delivery. For example, an analytic that characterizes the average cost of urban delivery in comparison to the number of deliveries would shed light on the efficiency gain related to automated deliveries. Another analytic that characterizes the average time for end-to-end delivery, taking account of the volume of deliveries, would also provide insight into efficiency gains. Improvement in user experience could be measured by a freight and logistics user satisfaction index and a freight management satisfaction index. These would measure the increased levels of satisfaction from the end user and from the freight operator, respectively. It is likely that such analytics will be closely related to and used in combination with transportation management analytics that characterize trip time and trip time reliability across the city. Ultimately this could support a more sophisticated approach to money-back guarantees for failure to deliver on time. This might even be extended to address mobility and transit services.

User-Focused Mobility

User-focused mobility services will make use of many of the analytics previously defined for the other 15 services. This would include a citywide mobility index to measure the increase in mobility caused by the service, a user satisfaction index to measure user perception of mobility services, and a reliability index for transportation services within the smart city.

An ultimate analytics for user-focused mobility services would compare the level of mobility afforded compared to the proportion of the population serviced and the resources invested in capital and operations.

6.7   Analytical Performance Management for a Smart City

As discussed in Section 6.5, the use of analytics will go beyond reporting to provide smart city managers with the ability to influence the performance of the city from a transportation perspective. This could be viewed as an extension to existing approaches to performance management and transportation. The use of analytics not only adds an extra dimension but also reinforces the need to consider the entire performance management process and not simply measurement. Analytics places a focus on the conversion of data to information and the use of information to create actionable strategies and insights. The use of analytics provides an extra layer in the analysis process of big data in the smart city. Analytics differ from KPIs or performance measures that are typically used for transportation service evaluation.

In the field of performance management, the term KPI is used to describe the parameters or data that are collected in order to measure performance. As discussed earlier, that is the difference between an analytic and a key performance indicator. Figure 6.3 [2] shows a list of KPIs from a European report that was designed to provide input into a cooperative framework for the application of advanced technology to transportation. It should be noted that the European experience with respect to analytics-driven performance management for transportation is slightly ahead of that of the United States at this point.

Note that each of the seven KPIs focuses on one aspect of transportation data. In comparison, an analytic typically draws on multiple data items and creates the relationship between them. For example, the analytic equivalent of KPI N1 would combine the change in peak period journey time data with additional data regarding the investment in intelligent transportation systems along the corridor to provide a characterization of the effectiveness of the investment. The analytic would be the percentage change in peak hour journey time per dollar invested in intelligent transportation systems along the corridor. Note that Figure 6.3 contains both long-list and short-list KPIs. The long list of KPIs was the result of a series of interviews with transportation stakeholders in European Union member states and industry experts. The short list represents an amended version of these KPIs, taking into account the inputs provided during a stakeholder workshop.

In fact, the name KPI provides insight into its use. The last word in the expression, indicator, conveys that these parameters are designed to indicate performance rather than provide insight into trends, patterns, and relationships. While analytics are not a substitute for KPIs or performance measures, they are a valuable addition to the tools that we can use for performance management in transportation and the smart city.

ID	Longlist KPI	Shortlist KPI
N1	Change in peak hour journey time in conjunction with flow between key points along a route (all vehicles)	% change in peak period journey time along routes where ITS has been implemented. Report by vehicle type where possible.
N2	Change in peak hour flow between key points along a route (all vehicles)	% change in peak period traffic flow along routes where ITS has been implemented. Report by vehicle type where possible.
N4	Journey time variability as measured using standard deviation of journey times between key points along a route (all vehicles)	% change in journey time variability on routes where ITS has been implemented - as measured by coefficient of variation. Report by vehicle type where possible.
N9	Modal shift (Change between personal cars and public transport)	% change in mode share on corridors where ITS has been implemented. Report percentage mode share separately for each mode where possible.
S1	Change in number of all reported accidents per vehicle km Change in severity of accidents (i.e. numbers killed or serious injured) per number of accidents reported)	% change in number of reported accidents along routes where ITS has been implemented. Report by accident severity (i.e. fatal, serious injury, light injury) where possible.
E1	Change in CO_2 emissions per vehicle km	% change in annual CO_2 emissions (Tons) on routes where ITS has been implemented.
L9	Number of automatically initiated eCalls	Time taken between initiation of public (112) eCall to the presentation of the content of MSD in an intelligible way at the operator's desk in the Public Safety Answering Point.

Figure 6.3   List of KPIs [2].

6.8   How Do Analytics and Data Lakes Fit Together?

On the basis that analytics tend to show relationships between different data elements, it is likely that greater insights will be revealed from more complete data sets. In this respect, analytics and data lakes fit together. The data lake is a rich set of data that has been drawn together across all transportation modes and transportation services within the smart city. Therefore, the data lake provides the source material for the use of analytics. The smart city transportation data lake will go beyond the traditional transportation data warehouse and include data elements that could be viewed as nontraditional, or data that would be stored in a separate location. For example, data regarding proposed investment plans and work programs would be combined in the data lake along with other items such as energy demand and retail transactions. A full-fledged, multisource data lake will not be created overnight but will evolve from supporting a few important primary use cases into a wider capability as other data is added, and the capability of the data lake is extended. One of the important elements of a data lake is the ability to draw data together from across an organization and an enterprise and create an enterprise-wide view of the data. In many transportation agencies, data is collected in silos and may even be stored for the use of individual staff members. This type of fragmented stovepipe data storage makes it very difficult to get the best value for the money from analytics. Fortunately, however, it is possible to create a virtual data lake, in which data resides in its original location but is indexed and accessible to the central repository. Chapter 9 details this subject.

6.9   How to Identify Data Needs Associated with Analytics

There is a chicken and egg problem that must be addressed with respect to the use of analytics. What comes first—the data or the analytic? The answer is typically that the analytics will be created first of all on the basis of needs that have been identified. However, it is not possible to conduct an analytic when the data required is not available. Therefore, in practice the initial list of data analytics would be filtered to ensure that the early analytics have the required data available. It is also the case that objectives are linked to use cases, which are linked to analytics. The best approach would be to pick one or two use cases that deliver clear and immediate value to the end user and for which any necessary data is available. The initial pass on the analytics should deliver value to the user through the use of actionable insights. The results of the initial analytics should also provide the business justification for further investments of time and money. It may also be the case during the initial analytics that ways in which the results can be improved through better data will be revealed. This sets the scene for the development of a structured data acquisition and use plan.

6.10   Summary

This chapter discusses the nature and characteristics of analytics from a transportation perspective. It presents a formal definition of the term analytic and interprets it for use in transportation and smart cities. In addition, the chapter addresses the difference between reporting, analytics, and KPIs, illustrating the point that all three are intended to work together for a smart city, while analytics reveal detailed insights into trends and patterns and create the basis for actionable insights that can influence the performance of smart city transportation services. Moreover, the chapter discusses the value of analytics in terms of its role in unlocking the value of big data. As big data grows in volume and variety and is collected at higher velocity, the value of analytics grows as a means to manage the sheer volume of data and turn it into meaningful information and insights. Accordingly, the chapter also discusses the progression from data to information to actionable insights. To show the relevance of analytics to transportation in a smart city, the smart city services defined in the previous chapter are assigned specific analytics. This illustrates the analytics that could be used to characterize the performance of service delivery under the 16 headings.

These analytics could be used as a starting point for the development of a specific set of transportation data analytics customized to the smart city. The chapter also discusses the use of analytics for transportation performance management within the smart city, including a direct comparison of KPIs from a European project to analytics. This is intended to emphasize the difference between key performance indicators and analytics. The chapter concludes with some information on how analytics and data links fit together, illustrating the symbiotic relationship between the two. The chapter also discusses the evolution of a data lake for smart city transportation, introducing the concept of early analytics work with carefully selected use cases, leading to business justification and further enhancement of the data lake.

Finally, the chapter introduces the identification of data needs for analytics, including the concept of early analytics work leading to further revisions and additional data collection. Analytics have the power to inform smart city managers and professionals from all aspects of transportation service delivery. To attain this, it is necessary to construct a bridge between data science and transportation. The middle ground role that interfaces between these two vital subject areas, will be both important and lucrative.

References

[1]   *Oxford English Dictionary,* http://www.oed.com/, retrieved October 16, 2016.

[2]   Study on Key Performance Indicators for Intelligent Transport Systems, Final Report, February 2015, AECOM LTD: http://ec.europa.eu/transport/sites/transport/files/themes/its/studies/doc/its-kpi-final_report_v7_4.pdf, retrieved Sunday, October 16, 2016.

7

The Practical Application of Analytics to Transportation

7.1 Informational Objectives of This Chapter

This chapter answers the following questions:

- What would make a good departure point for a smart city program?
- How would analytics be applied in integrated payment systems, Mobility as a Service (MaaS), traffic management, transit management, and performance management?
- What would make a good departure point for a smart city program?
- What are integrated payment systems, MaaS, traffic management, transit management, and performance management?
- What analytics should be applied to integrated payment, mobility service, traffic management, transit management, and to performance management?
- What services are enabled by integrated payment, MaaS, traffic management, transit management, and performance management?
- How should analytics be applied?

7.2   Chapter Word Cloud

Figure 7.1 presents a word cloud to provide an analysis of this chapter's contents, at a glance. The word cloud captures the most frequently used words within the document, with the font of each word in proportion to the frequency of use.

7.3   Introduction

This chapter explains the application of the analytics defined in Chapter 6. At this time, there are few examples of the use of analytics in U.S. cities. Therefore, the number of examples shown is limited; it, nevertheless, provides considerable insight into the practical application of analytics from a smart city transportation perspective. The objective is to build on Chapter 6 by explaining how analytics can be applied within a practical context in a transportation setting. To best explain the practical application of analytics to transportation, it would seem reasonable to use a series of examples.

During recent assignments on smart cities, it has become apparent that an important element of a smart city approach plan lies in the choice of the departure point. In this case the departure point is defined as a starting point for the proposed smart city initiative. If we assume that a smart city initiative will consist of the current situation, the implementation plan, and the ultimate smart city deployment, then it can be thought of as a journey, a departure point, a sequence of en-route activities, and a destination point.

Figure 7.1   Word cloud for Chapter 7.

It is likely that all cities will have a similar destination point in mind, assuming, of course, that we can all agree on the definition of a smart city. However, the departure point and the en-route activities will vary considerably depending on the current situation within the city and previous investments in advanced technologies. Therefore, there is an opportunity in this chapter to address two objectives simultaneously. The initial objective, and still the primary goal, is to explain the practical application of analytics within a transportation environment. The secondary goal is to explain a selection of departure points and make use of these as the examples for the application of the analytics. Five departure points are listed as follows, as examples for the practical application of analytics:

- Integrated payment systems;
- MaaS;
- Traffic management;
- Transit management;
- Performance management.

These departure points have been selected from the services introduced and defined in Chapter 5. As discussed in Chapter 5, the most robust approach to defining a smart city initiative is to make use of services that deliver value and benefits to citizens, visitors, and transportation service delivery organizations. For consistency, a selection of services has been identified as the starting or departure point for a smart city initiative, enabling the practical application of analytics to be illustrated.

The departure points are discussed in Sections 7.4–7.18, with each section explaining the major elements of the departure point as well as why it would make a good departure point for a smart city initiative. These explanations go deeper than the descriptions provided in Chapter 5 where the concept of services for a smart city are introduced Tables 7.1–7.5 list some candidate analytics that can be used in measuring the effectiveness of the departure points, along with some notes on the practical application of the analytics.

7.4 Integrated Payment Systems—What Are They?

An integrated payment system for a smart city would consist of the following elements:

- Electronic toll collection;
- Electronic ticketing for transit;

• Electronic fee collection for car parking.

An integrated payment system would incorporate all three to enable a single account to be used by a traveler to pay for all three services. While a single account would be used, it would be possible for the traveler to use one of a selection of defined payment instruments, such as a transponder for electronic toll collection, a smart card or a smart phone for transit ticketing, or fee collection for car parking. The single account for tolls, ticketing, and parking payments would be administered by an integrated back office capable of supporting transaction processing, customer service, and analytics.

The integrated back office could be achieved by the deployment of a single system or through conductivity between separate electronic toll collection, electronic ticketing for transit, and electronic fee collection for car parking systems. While a single integrated system may provide more benefits, the best use of sunk investment in legacy systems will probably be achieved through the development of middleware and interfaces between existing single-purpose systems.

7.5   Why Does Integrated Payment Make a Good Departure Point for a Smart City?

Many cities around the world have a significant investment in electronic payment systems of one form or another. These include electronic toll collection, electronic transit ticketing and electronic fee collection for car parking. This legacy of sunk investment in electronic payment systems is one reason why it would make a good departure point for a smart city. It is not just the dollars involved; it is also the acumen and experience that has been accumulated over the course of the investment in electronic payment systems.

Another reason for starting with an integrated payment system is an obvious one—it can generate revenue. An electronic payment system provides the mechanism for transportation services to be paid for on a pay-as-you-go basis. This provides an alternative to common funding for transportation services through taxation. This mechanism also provides a pathway for the private sector to be involved in the development of projects and the delivery of services. The ability to pay provides a mechanism for a return on investment by the private sector in return for the capital and operating resources required to establish the service. It is highly likely that the private sector will seek to find opportunities where profit is possible and where mechanisms for payment also exist. This might be one of the reasons why traffic signal systems have never really been privatized, certainly not in the United States.

A third and perhaps less obvious reason for starting with an integrated payment system within a smart city initiative is that the connectivity required to support payment could also form the basis for connected citizen and connected visitor service. The communication channels—the back office and the smart apps used to enable electronic payment and make payment for services fast and convenient—also have the potential to support a two-way information exchange between the back office and the traveler. In a similar fashion, an integrated payment system could also support many aspects of the performance management data collection required for a smart city. The ability to know when and where a transaction occurred as well as how much the transaction was worth could form the basis for a performance management system by providing data regarding the demand for transportation and current or prevailing conditions on the transportation network.

With respect to performance management in a smart city, an integrated payment system can also enable the use of congestion pricing, variable tolling, and variable parking fees to manage the demand for transportation. The London congestion charge discussed in Chapter 5, for example, explains the use of a mandatory congestion charge to manage the demand for private car use in central London. In the United States, while this concept has been studied, the focus has been placed on the provision of additional service quality levels in return for a fee, rather than a mandatory charge. This includes the implementation of express lanes where the toll charged is varied to achieve a specific predefined level of service. There are numerous examples of the application of dynamic tolling to express lanes in the United States [1]. Similar techniques can be applied varying parking fees to affect the demand for parking within the smart city. It is also possible to apply the same techniques to transit tickets, although there are likely to be institutional and political barriers to the application of variable pricing to transit.

7.6 Integrated Payment System Analytics and Their Practical Application

Table 7.1 contains a sample of analytics that can be used for an integrated payment system.

The analytics shown in Table 7.1 are applicable to electronic toll collection and electronic transit ticketing.

7.7 MaaS—What Is It?

MaaS is a relatively new concept that has been the subject of considerable discussion within the context of a smart city approach. It recognizes the emer-

Table 7.1
Candidate Analytics for Integrated Payment

Candidate Analytics	Application Notes
Active accounts as a proportion of total accounts	This would enable inactive accounts to be highlighted and action taken to either encourage customers to become active or to cancel accounts, thus reducing cost and increasing revenue.
Cost to manage each account	This enables the cost of managing each account to be determined and form the basis for cost-reduction exercises.
Proportion of transactions supported by each payment channel	Provides an indication of the balance of the different payment channels being used, enabling capacity optimization and repricing of payment channels if necessary.
Cost per transaction for each payment channel	Supports evaluation of the cost of each transaction by payment channel, thus providing guidance to marketing and outreach to promote each channel.
Revenue losses as proportion of total enforcement expenditure	A measure of comparative efficiency of the enforcement process against other integrated payment systems, forming the basis for improvement in the efficiency of the enforcement process.
Expenditure on revenue and fraud control as a proportion of total revenue	Compares the total resources expended in enforcement to the total revenue of the integrated payment system, providing a yardstick for the efficiency and effectiveness of the enforcement process.
Reduction of fraud or leakage as a proportion of expenditure on revenue and fraud control	This relates the total expenditure on fraud control and enforcement to the effects achieved in terms of a reduction in fraud and violations.
Revenue per customer	Provides an indication of the size of the customer base and an assessment of whether revenues are coming from a small or large number of customers.
Cost per statement	Provide an assessment of the efficiency of the billing system.
Customer satisfaction index	This will form the basis for improvements to customer outreach, communications, and service quality levels
Highest volume origin destination pair	This and other trip pattern analytics would enable careful matching of capacity with demand.
Location of premium customers	Knowing the location of premium customers would enable service improvements to be focused on those customers providing the highest value and car parking fee collection

gence of Uber and Lyft and their ability to drive privately operated mobility services to customers. It could be argued that such companies have successfully addressed the challenge of ride-sharing, which has been the subject of much investment and planning within the public sector. The emergence of these private services is also likely to change user perception with respect to the business model used to deliver transportation services. For example, it is possible that a traveler could decide not to purchase a personal vehicle, but make use of Uber and Lyft services instead. This introduces the concept of MaaS, under the auspices of which a traveler will acquire transportation on an as needed and pay-as-you-go basis as a service. Payment for the service could be related to use, or it could be provided for a fixed monthly fee. This offers the traveler an option to pay the monthly fee, rather than invest the money into the acquisition of a personal vehicle. The whole concept of MaaS is still evolving but will probably include the following elements:

- *Uber- and Lyft-style on-demand vehicle services:* On-demand private car transportation services provided on a ride-sharing basis. Both Uber and Lyft match available drivers and vehicles to those requiring transportation, making use of an advanced technology platform [2, 3].
- *Demand-actuated transit systems:* These services are typically operated by the public sector and involve the use of technology to enable transit passengers to call for service on demand.
- *Fixed-route transit services:* These services are also typically operated by the public sector and involve the use of transit vehicles operating along predefined routes and operating to a fixed schedule.
- *Flexible-route transit services:* These are a hybrid of demand-actuated and fixed-route services where transit vehicles operating within a zone can be diverted according to the demand for transit.
- *Paratransit services:* A specialized form of demand-actuated transit designed for people that cannot use any of the above services. Users should be preregistered to make use of the service before being able to reserve services in advance. Such services typically should be reserved at least one day in advance.

In the future, these services could also incorporate the use of fleets of self-driving or autonomous vehicles.

A service portfolio including the above elements would be presented to the traveler via web and smart phone applications. Information would be provided regarding the availability, the estimated travel time, the reliability, and the cost of each of the service options, enabling the traveler to choose the most appropriate option for the situation. There may also be close cooperation between public and private service operations. For example, currently, in central Florida, there are several cities that work in close cooperation with Uber. When the paratransit service funded and operated by the city closes at the end of normal working hours, it is possible for a paratransit user to make use of Uber instead. The city provides a subsidy for the Uber fare.

7.8 Why Does MaaS Make a Good Departure Point for a Smart City?

MaaS is a relatively new term that encapsulates considerable potential for improving transportation service delivery within a smart city. Over the past few years, significant privately financed and operated transportation service alternatives have emerged in the form of Uber and Lyft. These are overlaid on publicly funded transit services to offer the possibility of creating a portfolio of options for the traveler. The development and communication of this portfolio has the possibility of influencing the traveler away from the private vehicle and perhaps ultimately influencing decisions on whether to acquire a vehicle or simply acquire transportation as a service.

Convenient ways to request transportation and to pay for it provide significant influence in traveler decision-making. As most cities around the world are battling with congestion and the surplus of demand for private car transportation compared to public transit, MaaS could be viewed as one of the solutions to this issue. Cities also struggle with a lack of available land and the amount of land that car parking requires. In fact, there is a significant investment in private cars that at any given time is sitting in a parking lot and not returning value. MaaS offers the possibility for flexible and dynamic matching of supply and demand as it fluctuates over time and space within a smart city.

7.9 MaaS Analytics and Their Practical Application

MaaS represents a combination of both public and private transportation services that can be offered to the traveler as a portfolio. Table 7.2 lists some candidate analytics for MaaS and provides notes on the practical application of the analytics.

7.10 Traffic Management—What Is It?

The traffic management element will encompass the management of freeways, arterials, and city streets. It is often the case that these elements are managed independently, but we will assume that in a smart city there will be coordinated management across them. Freeway management includes the use of infrastructure-based and probe vehicle data collection, the use of in-vehicle systems, and roadside dynamic message signs to communicate with drivers and decentralize dispatching of incident clearance and recovery resources. In most cases, closed-circuit TV cameras are also used to understand the nature of the incident to provide input to the selection of resources to be dispatched.

Table 7.2

Candidate Analytics for MaaS

Candidate Analytics	Application Notes
Number of services available	The number of services available as a yardstick of transportation service choice; can form the basis for adding or supplementing services.
Destination served	The number of services available is another dimension of the choice available to the traveler. This can be matched to the demand for transportation to make sure that mobility has been provided.
Total cost of each service	The absolute cost of each service enabling comparisons between services and comparisons of cost of travel between different zones.
Cost of each service as a proportion of total household income	Enables the measurement of the affordability of each service and can be contrasted by trip and by zone.
Overall reliability of all services	Provides an indication of the quality of the service being delivered and could form the basis for strategies to improve reliability through investment in additional vehicles or advanced asset management techniques.
Reliability of each service	Focuses on the reliability of individual services as an indication of how desirable each service is.
Availability of the service	Availability can be measured in terms of routine availability and events that cause a reduction in the availability for each service.
Overall availability of all services	A measure of the overall level of availability for the service portfolio, hence providing a measure of service for the zone or traveler.

This will also include coordination with emergency services. Arterial traffic management involves the use of advanced traffic signal control to manage traffic at intersections. Sensors are used to measure traffic volumes, and sophisticated software programs are used to adjust signal timings to align with changes in traffic demand and to ensure coordination between adjacent intersections. The same techniques and technologies are applied to urban surface streets, where intersections are typically grouped more closely, and in many cases, the road pattern forms a grid with multiple traffic flows at intersections, in contrast to the corridor-like context of arterials.

For all three of these elements associated with advanced traffic management, the objective is to minimize congestion and delays by ensuring the smooth operation of freeways, arterials, and surface streets. Operation can be considered to take place under both recurring and nonrecurring congestion conditions. Nonrecurring congestion would be that associated with an incident, roadwork, or a special event. Recurring congestion is associated with the typical daily commute pattern and is usually caused by an excess of demand overcapacity at certain times of the day. Traffic management presents both a geographical and temporal challenge that can be very effectively addressed by analytics.

7.11   Why Does Traffic Management Make a Good Departure Point for a Smart City?

Traffic management is a good departure point for a smart city, not just because of the considerable sunk investment in transportation management systems within most urban areas, but also because of the high level of impact the traffic management systems can have on the smooth operation of city streets. Traffic management could be viewed as one of the original applications of advanced technology to transportation with electrical signal systems implementation dating back to the 1930s and earlier in Europe. Traffic management systems also provide one of the most visible applications of advanced technology that have a direct impact on the traveler.

Traffic management infrastructure, dynamic message signs, and traffic signals are highly visible in a very clear sign that an investment is being made to improve the transportation situation and the quality of the service delivery. Traffic management systems also benefit from a large base of expertise and experience gained over several years of implementation. There are many experts available in the field of traffic operations and traffic engineering that can support traffic management as an essential early element within a smart city initiative. Another important reason that traffic management is a good departure point for a smart city initiative lies in the visibility to the traveler. Major components of traffic management, including dynamic message signs, roadside sensors, and other roadside infrastructure, are extremely visible to the driver. It also doesn't hurt that a traffic management center provides an ideal photo opportunity for a politician.

7.12   Traffic Management Analytics and Their Practical Application

Table 7.3 lists a range of services that can be provided by traffic management along with some candidate analytics and some notes on how the analytics can be applied in a practical context.

7.13   Transit Management—What Is It?

Transit management entails the following activities:

- *Transit fleet management:* The use of advanced location technologies to develop a picture of the transit fleet at any given time. This typically captures the location of the vehicle and the vehicle identification.

Table 7.3

Candidate Analytics for Traffic Management

Candidate Analytics	Application Notes
Number of stops per vehicle	This analytic would be used as part of an overall approach to minimize the number of stops that a vehicle encounters along the route. This would require the use of optimization software on a network-wide basis and data regarding the traffic conditions encountered by each individual vehicle, from connected vehicle sources.
Travel time per vehicle	As above this would be part of an overall approach to minimize travel time for vehicle and to minimize the variability of travel time. This would also make use of optimization software and connected vehicle data sources.
Total incident response time	Total incident response time would span the process from incident detection, through verification, to dispatch of resources, then incident clearance and include traffic management over the duration of the process. Based on an incident response time study, incident response procedures can be streamlined and the assignment of resources on a geographical basis could also be considered.
Recurring congestion	A detailed analysis of recurring congestion, typically caused by commuters with a regular travel pattern. The study would lead to recommendations for increasing capacity and for better management of the recurring traffic conditions.
Number of pedestrians, cycles, buses, freight vehicles and private cars move through an intersection per hour	This analytic supports the optimization of the performance of each intersection on a multimodal basis. This would involve the read timing of traffic signals for private cars, transit vehicles, pedestrians, and bicycles.
Pedestrian, circle, bus, freight vehicle and private car delays at intersections	This analytic supports the optimization of the performance of the citywide traffic network on a multimodal basis. Would involve the re timing of traffic signals for private cars, transit vehicles, pedestrians, and bicycles.
Number of stops per vehicle for each trip	Enables the use of connected vehicle and optimization software to determine the number of stops per vehicle for each trip.
Variability of travel time per vehicle for each trip	As above but would address the variability of travel time per vehicle for each trip.
Signal timing is designed to optimize the trajectory of each vehicle through the intersection	Makes use of connected vehicle data to understand individual vehicle trajectories in in terms of instantaneous vehicle speed and stops. Signal timing will be optimized in the light of this new information to attempt to smooth traffic flow by reducing acceleration and deceleration due to stops.
Coordinated signal timings designed to optimize the trajectory of each vehicle through the network	As above but on a network-wide rather than individual intersection basis.

- *Passenger information:* The use of information technologies such as the web and smart phones to provide information to passengers regarding choices, service availability, service reliability, and support for payment of services. These include at-home services for trip-planning purposes and on-trip services to provide information regarding the next bus and route choices.

- *Transit ticketing:* Support for payment services can come from smart card–based electronic ticketing systems or more recently using Google pay or Apple pay, which both take advantage of host card emulation (using your phone as a credit card) and near field communications (using your phone to communicate with nearby payment devices) technologies. Note that with respect to transit ticketing, this departure point overlaps with the integrated payment.

7.14 Why Does Transit Management Make a Good Departure Point for a Smart City?

Transit management is a good departure point for a smart city for one simple reason: It can have a direct impact on the number of people that use transit. There are some cities around the world where the proportion of trips carried by public transit can be as high as 70% of the total trips made, but this is not the case in U.S. cities. In the United States, the total trips carried by public transit can vary from as little as 5% to more than 50%, leaving considerable scope for influencing the traveler toward using transit as a mode of travel. The efficiency of public transit is beyond question; for example, it is estimated that a 40-passenger bus could take off the road 40 private cars [4].

Travel by public transit is also cost-effective, space-efficient, and fuel-efficient. Transit management, with a focus on the most effective and efficient fleet management, combined with decision-quality information to passengers, has a significant potential to raise the market share for a smart cities transit system. Transit vehicles may also represent an early pathway for vehicle automation and improvement to vehicle fuel consumption and emissions.

In general transit management makes a good departure point, because in most cities there is a strong need to increase the ridership for public transit. This is particularly the case in most U.S. cities with a few exceptions (Washington, D.C., New York, and Boston come to mind). Most cities around the world have much higher public transit usage than U.S. cities, but it is still important to ensure proper usage and proper investment in public transit and to make sure that travelers are aware of the options and opportunities afforded to them.

7.15   Transit Management Analytics and Their Practical Application

Transit management involves the application of advanced technologies to transit fleet management, the delivery of passenger information, and the support for electronic payment services for transit. Table 7.4 captures candidate analytics that can be used for transit management applications. Table 7.4 also contains a column that provides a brief overview of how these analytics can be used in a practical situation.

7.16   Performance Management—What Is It?

Performance management involves the measurement of performance parameters for various aspects of transportation service delivery, followed by the development of insight and understanding based on these measures. The overall objective is to improve transportation service delivery based on a detailed understanding of how things work and prevailing operating conditions. A comprehensive approach to performance management would address each stage of the transportation delivery process, listed as follows:

Table 7.4
Candidate Analytics for Transit Management

Candidate Analytics	Application Notes
Travel times for each passenger	Makes use of movement analytics; travel times for each passenger on the transit network are measured and analyzed.
Travel time variability for each passenger	As above focusing on travel time variability for each passenger.
Bus utilization	This analytic can include the number of passengers per bus and the miles traveled by the bus, so the hours of service and support service optimization.
Revenue per bus	This analytic makes use of data from the integrated payment system to determine revenue per passenger in revenue per bus.
Passenger satisfaction index	Makes use of social media analysis of passenger satisfaction; an index can be developed that characterizes customer perception of service levels.
Revenue per passenger	Uses integrated payment system and movement analytics data to determine the revenue per passenger for service and payment structure optimization.
Revenue per route	Uses integrated payment system data and revenue per route service and payment structure optimization.
Comparison between schedule data and actual performance data	Schedule variation analytic is determined by comparing the scheduled to the actual performance as a measure of system reliability.

- Planning;
- Design;
- Project delivery;
- Operations;
- Maintenance.

Within each stage of the transportation delivery process, a complete performance management approach would include measurement of data, the conversion of data into information, the development of insight and understanding, and the definition of response and reaction strategies based on the insight and understanding. This is illustrated in Figure 7.2.

In the first step, data is collected that characterizes the performance of the mode or modes under consideration. In the second step, the collected data is converted into information through the aggregation and summarization of the data. In the third step, insight and understanding is developed based on the information provided from the previous step. This insight and understanding could include the identification of new connections and mechanisms. In the last step, strategies based on the insight and understanding developed in the previous step are implemented, and the effectiveness of the strategies is monitored using data collection and the rest of the process.

Historically, performance management has been addressed on a mode-specific basis with freeway management, traffic signals, transit, and freight all addressed separately. With the advent of smart cities and integrated corridor approaches, there is a new awareness of the need to coordinate and integrate

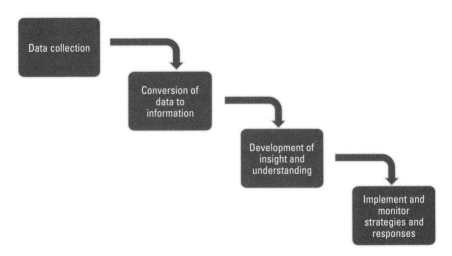

Figure 7.2   The performance management process for transportation.

performance management across all modes and across large geographic areas. The current state of the art of performance management in the United States focuses on measurement with less resources being placed on the other steps in the chain. However, money spent on data collection will not necessarily result in benefits unless the entire chain is addressed. Accordingly, it is vital to monetize the effort involved in performance measurement and ensure that it leads to actionable strategies for measurable improvements in the way that the enterprise operates, rather than just individual departments.

7.17 Why Does Performance Management Make a Good Departure Point for a Smart City?

As the old management adage goes, "If you cannot measure it then you cannot manage it." Performance management makes a good departure point for a smart city because of the ability provided to transportation service providers and city managers to measure and manage transportation service delivery. These days we would add something to the adage: "If you cannot measure it, you cannot manage it, and if you're only measuring it, you are still not managing it." A critical point in performance management is to make a difference, and just collecting data and reporting the information does not make a difference unless you take it the whole way toward action, strategy application, and reaction based on the new insights.

A comprehensive approach to performance management within a smart city can achieve these objectives and provide the basis for insight and understanding that will guide the establishment and operation of smart city transportation services in the most cost-effective and efficient manner. One of the critical reasons that performance management is important as a departure point for a smart city initiative is that it provides the ability to build a business case for investment. This includes sunk investment and future planned investment. The performance management system also provides the basis for ensuring that operations and service delivery are optimized, and every dollar provided for these is clearly subjected to high-quality stewardship.

7.18 Performance Management Analytics and Their Practical Application

Table 7.5 shows a sample of services that can be supported by performance management. Table 7.5 also lists a range of analytics that could be used, along with some notes on the practical application of those analytics.

Table 7.5
Candidate Analytics for Performance Management

Candidate Analytics	Application Notes
Ease or resistance to travel from home to health, education, and employment opportunities	Uses movement analytics (smart phone location) and transit schedule data to determine the ease or resistance to travel from essentially residential zones to those zones containing health, education, and employment opportunities.
Effectiveness of investments and matching of investment to problem locations	Uses movement analytics and work program data to analyze the effectiveness of investments by comparing before and after characteristics and considering the match between the location of the investments and the location of transportation problems
Optimizing the transportation system citywide to minimize accidents	Uses a combination of crash statistics to characterize the current situation and then develop a range of strategies designed to optimize citywide transportation service delivery from a safety perspective.
Adjusting the different modes of transportation to maximize conductivity required	As above but placing a focus on the transfer time or connectivity of different modes.
Price analysis by market segment and time of day. Price per passenger, price per origin/destination pairs. Value proposition for each passenger and each trip	Uses movement analytics and mode price data to compare the value proposition offered by different modes.

7.19   Summary

This chapter provides some information on the practical application of analytics to transportation. To provide the context for the application of the analytics, the chapter first identifies a series of departure points for a smart city initiative. These are explained in terms of the elements that comprise the departure point initiatives and the services that can be delivered by each departure point. For each departure point, a small selection of candidate analytics are defined and some advice is provided on the practical application of these analytics to improving service quality levels for transportation. Note that these are not the only possible departure points and that this is not intended to be a catalog but rather a practical exposition of how to apply analytics to transportation within a smart city environment. Nevertheless, the departure points included are ones that are likely to be attractive to a city when considering the needs of a smart city initiative.

Each smart city initiative will likely develop its own set of analytics based on specific needs. It is further likely that the early analytics for pilot projects and for proof of concept in the development of business case justifications will consider the value proposition that the use of each set of analytics will unleash and the availability of suitable data to support at least the first wave of analysis. This is likely to be an iterative process with lessons learned and practical experience from the early analytics providing input and recommendations for the improvement of data acquisition and data collection for subsequent analysis. This chapter intends to build on the information in Chapter 6 by providing some practical insight into how analytics can be defined and applied for transportation service delivery within a smart city and initiative or context.

It should also be noted that some analytics will be discovered during early analytics work. The nature of the analytics process allows for a certain amount of discovery, and the latest approach to data and information management is to find ways for the data to speak. A skilled analyst working with a combination of data sets within a data lake is highly likely to uncover some new relationships that will lead to the definition of new analytics. Therefore, it is probable that a range of predefined analytics will be used to support the initial analytics work and that these will be supplemented with additional analytics that are discovered during the analytics work. This provides the potential for breakthroughs in understanding and insight as new analytics, new connections, and new mechanisms are identified and defined because of the combined data set or data lake. This feature has the scope for considerable innovation on the interface between transportation and data science and is likely to be a fertile subject area for research and the practical application of analytics to transportation.

The application of analytics is likely to take place within a wider context of evaluation and understanding the effects of transportation investments. It is useful to consider analytics, along with the data lake, as tools that work very well together, with the availability of data enabling a rich set of analytics.

As a final note in this chapter, it is also worth considering the future role of analytics within a path toward total automation of the back office. In fact, the ease with which data can be sourced for the purposes of smart city transportation analytics is significantly impacted by the formation of a data lake. Currently, data is often kept in a fragmented, poorly cataloged form. Merging data and having an organization-wide view of available data is an important step in enabling the analytics discussed in this chapter.

There is considerable effort and interest in the concept of autonomous vehicles, and this could be considered as the application of automation to one component of the overall vehicle and the highway infrastructure. It is reasonable to assume that the back-office component of transportation systems will also be subject to the same level of automation. Today, we have transportation management center operators and managers reviewing and evaluating data regarding current transportation conditions, in some cases with the help of sophisticated decision-support systems.

Future traffic management systems could feature a higher level of automation, based on a more detailed understanding of causes and effects. This defines the role for analytics in the future transportation system. We can start now by identifying analytics that characterize transportation conditions both now and in the future and develop a better understanding of cause and effect. This can form the basis for artificial intelligence and machine learning that can pave the way for more automation.

References

[1] U.S. Department of Transportation, Federal Highway Administration, Office of Operations, "Managed Lanes: a Primer," https://ops.fhwa.dot.gov/publications/managelanes_primer/, retrieved April 8, 2017.

[2] How Does Uber work?, https://help.uber.com/h/738d1ff7-5fe0-4383-b34c-4a2480efd71e, retrieved November 13, 2016.

[3] How Lyft works, http://blog.lyft.com/posts/how-does-lyft-work retrieved November 13, 2016.

[4] Host Card Emulation (HCE) 101, Smart Card Alliance, mobile, and NAC counsel white paper, August 2014, available at http://www.smartcardalliance.org/downloads/HCE-101-WP-FINAL-081114-clean.pdf, retrieved November 27, 2016.

8

Transportation Use Cases

8.1 Informational Objectives of This Chapter

This chapter answers the following questions:

- What is a use case?
- What information will it capture?
- What use cases could be valuable in a smart city transportation context?
- How can use cases be applied?
- How can use cases be related to smart city services as defined in Chapter 5?

8.2 Chapter Word Cloud

A typical way to visualize a set of data or document an overview is to create a word cloud, a diagram that lists all the most frequently used words within the document, sizing the font of each word in proportion to its frequency of use. Figure 8.1 shows the word cloud for this chapter.

Figure 8.1   Word cloud for Chapter 8.

8.3   Introduction

On first approach, the term use case can be daunting. However, it is a term that is a widely used in software and system engineering. Use cases have also been adopted in big data and analytics practice, making it worthwhile to become familiar with the term. Furthermore, use cases can be used in effectively in smart city planning and implementation to guide the application of big data and analytics techniques. This chapter provides an overview of the use case as applied to smart cities. The intention is to explain the definition of a use case, illustrate how use cases are applied in practice, and provide some examples of smart city transportation use cases. These examples are not intended to be a complete or comprehensive catalog of all use cases that could be identified and defined for a smart city. Specific use cases that are applicable to a smart city will vary according to smart city needs and the smart city departure point and will be customized to individual smart city initiatives.

The application of big data and analytics to transportation within the city is a relatively new area, and it is hoped that the provision of these illustrations and examples will stimulate detailed thinking that will lead to the definition of a wider range of use cases. This chapter adopts a standard format for the documentation of use cases. It does not follow the rigorous system or software engineering format but is tailored to the needs of smart city transportation services. The focus is on linking the objectives or questions to be answered with the analytics that support the attainment of objectives. Sixteen smart city transportation use cases have been identified, described, and explained. Each use case is related to an initial set of objectives, and the intended user of the analytics will be defined and described. A high-level set of intended results or benefits is also identified.

8.4   What Is a Use Case?

The term use case means different things to different people, depending on what they are trying to achieve using the use case.

According to Wikipedia [1], a use case is defined as:

> …a list of actions or event steps, typically defining the interactions between a role (known in the Unified Modeling Language as an actor) and a system, to achieve a goal. The actor can be a human or other external system. In systems engineering, use cases are used at a higher level than within software engineering, often representing missions or stakeholder goals. The detailed requirements may then be captured in the Systems Modeling Language (Sims) or as contractual statements.

This is a software and system engineering definition. In the world of software and system engineering the use case has a prescribed format and is comprised of rigorous definitions of data flows, actions, events, and interactions between people and systems. In this case, they form the basis for system and software design requiring such rigor and detail.

The purpose of the use case within the context of this book is as a significant component in the bridge between smart city needs and data science capabilities. An understanding of the term can be derived by decomposing it into two elements: use and case. *Use* indicates that it describes the application or use of the proposed analytics. It also indicates that the end result should be useful. *Case* indicates that it is an example or an illustration of how the analytic can be applied. Putting the two together yields an explanation of the use case as an illustration of how the analytics will be applied and how they will deliver results. This, in the simplest of terms, is the function of the use case as defined for the purposes of this book. For the purposes of applying big data and analytics to smart cities, a simplified, less rigorous use case definition is appropriate. The objective is not to support software or system engineering design, but to act as a bridge and as a communication tool between smart city transportation experts and data science experts.

The identification and definition of suitable analytics for smart cities requires an understanding of the transportation needs and services to be supported combined with an understanding of the capabilities of data science and analytics. Smart city transportation use cases are designed to capture things that can be done, the objectives of the analysis, and the data needs. They are used to explain to the end user that needs and objectives have been understood and addressed and that the value proposition of conducting the analyses has been thought through and documented, at least in outline. The use case sets the scene for the conduct of the analytics work, acting as a major guiding force to ensure that the analytics are firmly focused on objectives and business value. An initial overview of data requirements also ensures that use cases are developed with full recognition of available data. Use cases are used to link analytics to objectives. They also serve as a communication tool in two directions. In the first direction, objectives are linked to analytics ensuring that end users can see how their objectives are being matched. This could be thought of as feedback. Use cases also support communications in the other direction, which could be considered to be feed-forward. The use case communicates the problem and details of the proposed data and analytics to the data scientists or analyst.

8.5   Smart City Transportation Use Case Examples

Appendix A identifies and describes a collection of 17 smart city transportation use case examples. Table 8.1 provides an overview of the use cases and the smart city transportation services to which they relate.

Each use case example in Appendix A follows the same format, with the following elements:

- *Smart city service:* A description of the smart city service that is addressed by the use case. The use cases are directly connected to the 17 smart city services defined in Chapter 5.

- *Use case name:* A short label for the use case that reflects the subject area. The label is intended to make it easy to refer to the use case in a short-hand manner during the application.

- *Objective and problem statement:* A concise definition of the business challenge to be addressed by the use case. This is to ensure that user needs and objectives have been understood and captured.

- *Expected outcome of analysis:* A description of the outcome of the analysis that will deliver benefits. This ensures that the desired outcome of the analysis has been clearly defined at the outset.

- *Success criteria:* Critical success factors in the delivery of the use case. The support for and objectives-driven approach to analytics, reinforcing the focus on objectives and user needs, avoiding a disconnect between the analytics and the objectives.

- *Source data:* A high-level description of data content, data latency, data detail, and any further information regarding the nature of the data nec-

Table 8.1
Smart City Transportation Services and Use Cases

	Smart City Service	Use Case Name
1	Asset and maintenance management	Asset and maintenance management
2	Connected vehicle	Connected vehicle probe data
3	Connected involved citizens	Connected, involved citizens
4	Integrated electronic payment	Variable tolling
5	Integrated electronic payment	Ticketing strategy and payment channel evaluation
6	Intelligent sensor–based infrastructure	Intelligent sensor–based infrastructure
7	Low-cost, efficient, secure, and resilient ICT	ICT management
8	Smart grid roadway electrification and electric vehicles	Electric fleet management
9	Smart land use	Mobility hub
10	Strategic business models and partnering	Partnership management
11	Transportation governance	Transportation governance system
12	Transportation management	Customer satisfaction and travel response
13	Travel information	Travel value analysis
14	Urban analytics	Accessibility index
15	Urban automation	Urban automation analysis
16	Urban delivery and logistics	Freight performance management
17	User-focused mobility	MaaS

essary to ensure that data is available to support the use case and analytics. This will be an initial review of data requirements to avoid the development of use cases that require data that is not available. It is likely that data requirements will evolve as the analysis takes place.

- *Business benefits:* Key benefits to be delivered by the use case in terms of safety, efficiency, and user experience enhancement. These are stated as business benefits since a methodology was drawn from significant experience in working with private sector companies. These can translate into public benefits such as accident reduction and reduced travel time and improve travel time reliability and provide a better understanding of user experience and perception.

- *Challenges:* The identification of challenges that need to be addressed to ensure successful delivery of the use case. This represents an initial list of challenges that can be predefined before the analytics work takes place. These will be supplemented by additional challenges and covered during the work. This initial list is intended to support the early identification of challenges in the development of appropriate approach plans.

- *Analytics to be used:* A list of the proposed analytics that will form the basis of the output of the analysis work. Here again, this is an initial list of analytics to be used and will be supplemented by additional analytics that will be discovered during the work. The analytics can be supported by several different analytics techniques, including the following:
  - *Graph analytics:* Initial relationships between data elements and people; can also show the strength of the relationship based on data attributes.
  - *Text analytics:* These can uncover underlying sentiment within social media, and compliance are infractions and communications and documents of all kinds. They can also be word cloud visualizations as used in this book.
  - *Path pattern and time series analytics:* These provide insight on interaction patterns between people, products, or data elements.
  - *Structured query language (SQL):* This is a standardized query language for requesting information from a database. It provides flexible ways to manipulate data and to make queries from a big data set using the language of business tools.
  - *Statistical modeling:* This includes statistical modeling techniques such as linear least squares regression, nonlinear least squares regression, weighted least squares regression, and locally weighted scatterplot smoother (curve fitting).
  - *Machine learning:* Techniques to sift through data with minimal human input to gain new insights previously undetected. This can form the basis for decision support and automation.

This format is an approach adopted by a major big data and analytics practitioners and solution providers [2] with many years of experience in developing and implementing use case descriptions associated with big data and analytics projects.

The purpose of Appendix A is twofold: to explain a few transportation use cases and to illustrate how the use cases are put together in practice. The intention is to provide practical examples of use cases that can be applied to smart city transportation initiatives. These can be used as a starting point for a more complete set and as a model on how to create a practical use case template.

8.6   Summary

The use case is a very important tool to gauge the implementation of big data and analytics techniques regarding smart city transportation. This chapter explores the definition of the term use cases and explains a simplified version of the traditional use case that best serves the need for smart city transportation. The chapter also identifies and defines a sample of 17 smart city transportation use cases along with the type of information that would be captured in the use case and practice. These serve to illustrate the application of the use case technique and provide some examples of smart city transportation use cases. Furthermore, they are intended to stimulate thought on the definition and development of a wider range of smart city use cases. It is expected that smart city use cases will be customized for each smart city implementation. As the term use case can be difficult to grasp at first, this chapter explains the concept and how it can be adapted for use in smart city transportation service delivery. The ability to document user objectives and the value proposition to be delivered as a result of the analytics work is crucial to justifying the effort required in applying big data and analytics to the smart city transportation environment.

References

[1]   Use case definition, Wikipedia, https://en.wikipedia.org/wiki/Use_case#Templates, retrieved December 3, 2016.

[2]   Teradata use case template.

Appendix A: Smart City Transportation Use Case Examples

Use Case Example 1: Asset and Maintenance Management

Smart City Service: Asset And Maintenance Management

- *Objectives:* To improve the quality of asset and maintenance management, to minimize cost and maximize results, and to support a consistent and appropriate quality level for assets across the city.
- *Expected outcome of analyses:* Better cost versus performance results for asset and maintenance management, more consistent and appropriate levels of maintenance, better understanding of relevant intervention points and replacement.
- *Success criteria:* Better asset and maintenance management performance, improvement in asset performance, improved consistency of maintenance, and development of strategies for optimum asset and maintenance management.
- *Source data examples:* Asset location, asset condition, maintenance logs, maintenance schedules, maintenance service specifications and standards, maintenance program expenditure data, cost of individual device maintenance, cost of individual device replacement, cost of network maintenance, and cost of network replacement.
- *Business benefits:* Reduced costs, enhanced life cycle, managed maintenance costs, and better maintenance planning.
- *Challenges:* Establishing suitable maintenance standards, agreeing on maintenance standards across multiple responsible agencies, and developing an asset inventory.
- *Analytics that can be applied:* Establishing suitable maintenance standards, agreeing on maintenance standards across multiple responsible agencies, and developing an asset inventory

Use Case Example 2: Connected Vehicle Probe Data

Smart City Service: Connected Vehicle

- *Objectives:* To support maximum use of data that can emanate from connected vehicles and provide new data feeds that can be incorporated into existing ones; lessen the dependence on infrastructure-based sensors.
- *Expected outcome of analyses:* Significantly improved picture of transportation operating conditions and the demand for transportation in urban

areas; integrated use of probe vehicle and sensor-based data; and maximizing the value of the data delivered and minimizing the cost of data collection.

- *Success criteria:* Use of vehicle probe data for the full spectrum of transportation activities; effective integration of probe- and sensor based data; and optimization of data collection and acquisition investments.
- *Source data examples:* Connected vehicle data including vehicle location, instantaneous vehicle speed, vehicle ID, and vehicle dynamics and engine management data.
- Business benefits: More comprehensive and higher resolution picture of transportation supply conditions and transportation demand.
- *Challenges:* Agreeing on access to connected vehicle data and improving market penetration of connected vehicles.
- *Analytics that can be applied:* Connected vehicle data accessibilty, connected vehicle market penetration.

Use Case Example 3: Connected, Involved Citizens

Smart City Service: Connected, Involved citizens

- *Objectives:* To support a two-way dialogue between data sources and citizens and to enable citizens to provide crowdsource data and feedback concerning perception of quality and satisfaction levels.
- *Expected outcome of analyses:* Better informed citizens and enhanced abilities for citizens to provide data and opinions on transportation service delivery.
- *Success criteria:* Higher levels of citizen satisfaction and an increased awareness of citizen perception of traveler information service quality.
- *Source data examples:* Movement analytics data; citizen perception data; and quality of transportation service data.
- *Business benefits:* Enhanced user experience; increased understanding of user perception; and lower cost of data collection by incorporating crowdsourcing.
- *Challenges:* Developing a suitable data collection that can also enable user perception feedback and integrating user perception and crowdsourcing data with other data.

- *Analytics that can be applied:* User perception index, user perception comparative index.

Use Case Example 4: Variable Tolling

Smart City Service: Integrated Electronic Payment

- *Objectives:* To understand the relationship between levels of tolls and demand for use of toll roads. This is typically referred to as elasticity. The objective is to make use of observed data to improve the accuracy of descriptive and predictive elasticity analytics.
- *Expected outcome of analyses:* A more detailed understanding of toll elasticity, taking account of additional factors such as weather and trip purpose, and better predictions regarding the relationship between toll levels and demand for travel.
- *Success criteria:* Improved understanding of the relationship between toll levels and the demand for travel and toll roads.
- *Source data examples:* Toll transaction volumes, toll revenues, volume of traffic on each segment of the toll road, prevailing toll rates, trip purpose data, prevailing weather data, and origin and destination data.
- *Business benefits:* Maximized revenue through the successful application of a more detailed understanding of elasticity and improved user experience due to a stronger ability to preserve levels of service.
- *Challenges:* Trip purpose data could be a challenge that will require a creative approach.
- *Analytics that can be applied:* Revenue related to traffic flow and toll level related to traffic flow, taking into account weather and trip purpose factors.

Use Case Example 5: Ticketing Strategy and Payment Channel Evaluation

Smart City Service: Integrated Electronic Payment

- *Objectives:* To conduct what-if analyses on various ticketing strategies to identify the optimum strategy that will achieve objectives while offering the best value for money to travelers; to analyze the cost and value of different payment channels such as smart cards and host card emulation to determine the optimum balance between payment channel use; and to develop strategies for the achievement of the optimum payment channel use pattern.

- *Expected outcome of analyses:* Better use of payment channels and optimization of ticketing strategies that deliver greater efficiency and enhance the user.

- *Success criteria:* Maximizing revenue from each payment channel, compared to the cost of operating each channel and identifying and applying the most appropriate ticketing strategy for each mode.

- *Source data examples:* Volume of revenue from each payment channel; ticketing strategy data; and effects of strategies on revenue data.

- *Business benefits:* Lower cost of payment channel operation; enhanced user experience; and maximized revenue.

- *Challenges:* Collecting performance data on each payment channel and developing a catalog of possible ticketing strategies.

- *Analytics that can be applied:* Payment channel efficiency, ticketing strategy effectiveness, cost of money collection, relative use of each payment.

Use Case Example 6: Intelligent Sensor–Based Infrastructure

Smart City Service: Intelligent Sensor–Based Infrastructure

- *Objectives:* To optimize the balance between the cost of intelligent sensor–based infrastructure and the quality of the data delivered and to promote an integrated approach to data collection that blends together infrastructure-based sensors and probe vehicle sensors.

- *Expected outcome of analyses:* Better use of infrastructure-based and vehicle-based sensors in an integrated fashion.

- *Success criteria:* More efficient data collection, reduced cost of center operation, better integration of sensor data with other data.

- *Source data examples:* Sensor-based data including volumes and speeds; cost of data collection by different means; and quality of data by different means.

- *Business benefits:* Lower cost of data collection; better and more complete data; and better management of investments in infrastructure.

- *Challenges:* Establishing data quality targets; measuring data quality; and integrating data from multiple sources.

- *Analytics that can be applied:* Data quality index, sensor efficiency, cost-benefit to sensors.

Use Case Example 7: ICT Management

Smart City Service: Low-Cost, Efficient, Secure, and Resilient ICT

- *Objectives:* To obtain the best value for money from investment in information and communication technologies; to optimize operations of the communications network; to minimize the cost of data transfer; and to maximize network security.

- *Expected outcome of analyses:* Optimized use of information and communication technologies; better management of investments to achieve maximum results with minimum expenditure; and better use of private sector resources.

- *Success criteria:* Minimum cost communication services and optimized load-balancing.

- *Source data examples:* Cost of implementation and maintenance for information and communication technologies; network load data; and network demand data.

- *Business benefits:* Lower cost of network operation and higher standards of network operation.

- *Challenges:* Measuring network load; measuring network demand; and developing better utilization strategies for the network.

- *Analytics that can be applied:* Network load, network demand, network utilization, efficiency of network utilitzation strategies.

Use Case Example 8: Electric Fleet Management

Smart City Service: Smart Grid Roadway Electrification and Electric Vehicles

- *Objectives:* To support the analysis of electric vehicle charging point location patterns and the determination of energy use patterns associated with electric vehicles; to support a detailed analysis of energy use patterns; and to support the development of strategies for managing energy use.

- *Expected outcome of analyses:* Appropriate location of electric vehicle charging points to maximize availability to electric vehicle users and better matching of supply and demand from an energy perspective, taking account of the new energy consumption patterns related to electric vehicles.

- *Success criteria:* Availability of electric vehicle charging points; optimized energy consumption related to electric vehicles; and increased use of electric vehicles and the urban environment.

- *Source data examples:* Number of electric vehicles; energy and demand related to electric vehicles; population distribution; electric vehicle distribution; energy consumption data for electric vehicles; and electric vehicle use data.

- *Business benefits:* Energy efficiency; etter matching of energy supply and demand; reduced omissions; and reduced dependency on fossil-based fuels.

- *Challenges:* Determining demand for electric vehicle charging; collecting electric vehicle use data; and collecting electric vehicle ownership data.

- *Analytics that can be applied:* Demand for charging electric vehicle use, electric vehicle ownership, market penetration of electric vehicles, miles traveled for electric vehicles as compared to other vehicles.

Use Case Example 9: Mobility Hub

Smart City Service: Smart Land Use

- *Objectives:* The use of observed data to provide a detailed understanding of the relationship between land use and transportation demand.

- *Expected outcome of analyses:* More accurate and detailed assessment of the effects of land use on transportation demand.

- *Success criteria:* The development of better strategies for relating land use, transportation demand, and transportation supply.

- *Source data examples:* Origin and destination data; movement analytics data; smart manufacturing data; smart retail data; and smart healthcare data.

- *Business benefits:* Better decision-making data and deeper understanding of the relationship between land use and transportation demand.

- *Challenges:* Access to observe data; access to mobility analytics; characterizing existing land use; and developing a catalog of land use transportation impacts based on observe data.

- *Analytics that can be applied:* Mobility hub efficiency, mobility hub throughput, mobility hub cost-benefit index.

Use Case Example 10: Partnership Management

Smart City Service: Stragic Business Models and Partnering

- *Objectives:* To support the establishment and effective management of partnerships between the public and private sector with respect to smart city initiatives.

- *Expected outcome of analyses:* More effective public-private partnership management.

- *Success criteria:* Maximum leverage of private sector resources and achievement of public policy objectives with optimized investment.

- *Source data examples:* Partnership objectives data; costs related to the partnership; rewards related to the partnership; and transportation service delivery data.

- *Business benefits:* More effective partnerships; more sustainable partnerships; better leverage of private sector resources to achieve public sector objectives; and the development of sustainable new business enterprises within the smart city environment.

- *Challenges:* Collecting data on public-private partnerships; establishing objectives of public-private partnerships; and determining revenue data for public-private partnerships.

- *Analytics that can be applied:* Partnership efficiency, private sector investment levels.

Use Case Example 11: Transportation Governance System

Smart City Service: Transportation Governance

- *Objectives:* To support more effective transportation governance by providing the means to analyze the likely impacts and effects of various governance structures and to provide insight into the relationship between the delivery of smart city services and the required organization for success.

- *Expected outcome of analyses:* Governance structures are carefully aligned to the needs of transportation service delivery in a smart city.

- *Success criteria:* More effective governance related to transportation service delivery in urban areas.

- *Source data examples:* Transportation cost data; transportation benefits data; transportation supply data; transportation demand data; transportation coordination data; and data storage cost data.

- Business benefits: Better more effective governance; lower cost of governance; and better support for transportation service delivery in a smart city.

- *Challenges:* Developing a catalog of suitable government structures; measuring governance efficiency; and collecting supply and demand data.

- *Analytics that can be applied:* Efficiency of government structures, effectiveness of governance strategies, cost of governance as a proportion of total budge.

Use Case Example 12: Customer Satisfaction and Travel Response

Smart City Service: Transportation Management

- *Objectives:* To support a comprehensive approach to transportation management that includes insight into customer satisfaction and travel response as well as technical parameters related to transportation service delivery performance.

- *Expected outcome of analyses:* Transportation service delivery strategies that take account of user perception and satisfaction as well as technical performance parameters.

- *Success criteria:* Higher customer satisfaction with respect to transportation services and improved performance management that takes account of user perception in addition to actual performance measurements.

- *Source data examples:* Customer satisfaction data from smart app; actual transportation service delivery performance data; mobility data; jobs data; and travel time data.

- *Business benefits:* Enhanced user experience and improved performance management.

- *Challenges:* Collecting user satisfaction data,; collecting mobility data; and collecting accessibility data.

- *Analytics that can be applied:* User satisfaction index, mobility accessibility, model select.

Use Case Example 13: Travel Value Analysis

Smart City Service: Travel Information

- *Objectives:* The determination of traveler value for each mode and route, taking into account total trip time, trip time reliability, and cost of travel as a proportion of household income.
- *Expected outcome of analyses:* Ensuring equity for all travelers and optimizing transportation service delivery across the city.
- *Success criteria:* Improved social equity; improved transportation services; and better balancing of travel value across the city.
- *Source data examples:* Origin and destination data; trip travel time data trip cost data; and household income data.
- *Business benefits:* Better traveler decisions; improved equity; and better matching of transportation services to user needs.
- *Challenges:* Collecting traveler satisfaction and perception data and collecting behavior change data.
- *Analytics that can be applied:* Traveler satisfaction index, perception analytics, behavioral change analytics.

Use Case Example 14: Accessibility Index

Smart City Service: Urban Analytics

- *Objectives:* Determination of the ease or difficulty of travel from residential zones to job opportunities, healthcare, and education opportunities.
- *Expected outcome of analyses:* The configuration of transportation services to maximize accessibility to jobs, education, and healthcare.
- *Success criteria:* Improved accessibility to jobs, healthcare, and education.
- *Source data examples:* Origin and destination data; residential zone data; jobs on data; healthcare zone data; and education zone data.
- *Business benefits:* Enhanced accessibility achieved by better matching transportation needs to transportation service provision.
- *Challenges:* Access to job, healthcare, and education opportunity data; and access to residential zone demographics data.
- *Analytics that can be applied:* Job accessibility, healthcare accessibility, education accessibility, residential zone demographics.

Use Case Example 15: Urban Automation Analysis

Smart City Service: Urban Automation

- *Objectives:* Analysis of progress toward the application of urban automation, including the movement of people and goods.
- *Expected outcome of analyses:* Accelerating the deployment of automation within the urban environment.
- *Success criteria:* Accelerated progress in implementing urban automation.
- *Source data examples:* Automated vehicle use data and transportation demand data.
- *Business benefits:* Transportation service cost reduction; improved transportation service reliability; and better transportation service response.
- *Challenges:* Access to data on ownership and use of automated vehicles.
- *Analytics that can be applied:* Vehicle ownership data, automated vehicle use data.

Use Case Example 16: Freight Performance Management

Smart City Service: Urban Delivery and Logistics

- *Objectives:* Detailed assessment of the cost of urban delivery for goods, average time for entering delivery and quality of delivery service
- *Expected outcome of analyses:* More effective urban delivery for goods; better value for money for goods customers; and an increase in service quality.
- *Success criteria:* Lower-cost urban goods delivery; minimized cost of goods delivery; and maximized goods delivery service quality.
- *Source data examples:* Urban delivery cost data; urban delivery trip time data; user satisfaction data; and operator satisfaction data.
- *Business benefits:* Reduced freight cost; enhanced freight delivery time reliability; and enhanced user experience.
- *Challenges:* Access to freight delivery costs and access to delivery times.
- *Analytics that can be applied:* Freight delivery costs, delivery times, delivery time reliability.

Use Case Example 17: MaaS

Smart City Service: User-Focused Mobility

- *Objectives:* Optimization of transportation services to maximize mobility across the city.
- *Expected outcome of analyses:* Providing more flexible choices for mobility within an urban area including information regarding currently available options for mobility from both the public and private sector.
- *Success criteria:* Raised awareness of transportation service options and improved decision-making with respect to transportation choices.
- *Source data examples:* Origin and destination data and transportation service option data including availability, cost, and reliability.
- *Business benefits:* Increased mobility; transportation service cost reduction; enhanced user experience; and improved transportation service reliability.
- *Challenges:* Establishment of a mobility as a service portfolio; access to origin and destination data; and access to transportation service delivery data.
- *Analytics that can be applied:* Percentage of utilization of each transportation service, cost of transportation, travel time, travel time reliability.

9

Building a Data Lake

9.1 Informational Objectives

This chapter addresses the following informational objectives:

- It describes current fragmented approaches to data collection storage and management.
- It defines the data lake concept.
- It explains how a data lake works.
- It describes the key elements of a data lake.
- It explains the value of a data lake.
- It identifies some challenges that will be faced in the creation of a data lake, based on previous experience.
- It describes an approach to building a data lake.
- It explains the possibilities for organizational fine-tuning around the operation of a data lake.

9.2 Chapter Word Cloud

Figure 9.1 shows a word cloud for this chapter.

Figure 9.1   Chapter 9 word cloud.

9.3   Introduction

Chapters 2, 3, and 5–7 touch on data lakes, with a basic definition provided in Chapter 2 when the term was first introduced. This chapter provides a detailed explanation of the term data lake, an explanation of the value of the data lake and a suggested robust approach toward building a data lake. The data lake is an analogy for bringing data together from different sources, making it accessible and transforming it into a format that can be useful across the enterprise or organization. The analogy or visual offered by the term is an extremely useful communication tool, and it is valuable when introducing new data science and data analytics concepts to transportation professionals. Ideally, transportation specialists will be able to maintain a focus on their area of expertise, while making use of new data science and analytics tools to assist in gaining new insights and understandings. The data lake analogy allows the overall characteristics of the concept to be discussed and a value to be defined without diving into the weeds with respect to data science and data analytics. Bearing in mind that this book is designed to provide an overview of big data and data analytics techniques for smart cities, the data lake analogy provides an ideal communication tool. The purpose of this chapter is not to provide a how-to guide on selecting and using technology related to dig data and analytics. Rather, it is intended to provide an overview of how the data lake fits within the bigger picture and the value that can be delivered by taking this radically new approach to the storage of data. Providing an overview, rather than a detailed exposition, also avoids technology selection or bias toward a specific set of tools. While some specific solutions and approaches are used to illustrate the data lake concept, the overall approach allows the selection of multiple technologies and solutions to fit within the needs of the specific smart city.

There is a multitude of alternatives with respect to technologies that can be deployed to establish and maintain a data lake. However, the most important aspect associated with developing data lakes lies in taking a robust planning approach. Such an approach should take full advantage of previous experience and lessons learned to accommodate a flexible choice of technologies and solutions.

The emergence of large-scale data storage and manipulation technologies such as Hadoop [1] enables a new philosophy of data aggregation and consolidation into a single repository, rather than the earlier approach where data had to be divided and partitioned to make it manageable. A data lake is a virtual concept as it is feasible to allow data to remain in the existing source while making a copy available for use in the data lake. Again, this depends on the exact choice of the solution of technology to be deployed.

The contents of this chapter will be akin to a waterskiing adventure across the data lake, rather than a deep dive into specific technologies and products within data science and analytics.

As stated in Chapter 2, the data lake analogy is useful in as it suggests a clean or filtered body of water that contains useful and accessible data. It is not a data swamp, which would contain both useful and not so useful items in a mixture that would make the data less accessible. The data lake concept places an emphasis on bringing data together, making it accessible and visible across an organization or enterprise.

The creation of a data lake involves the removal of silos and partitions that are present because of the way the data has been collected managed and utilized in the past. Work assignments with several transportation agencies have revealed a natural tendency for data to be collected in what could be referred to as cockpits, with a cockpit being an array of data that is assembled by an individual or team with the objective of supporting a specific job function. For example, a traffic engineer might collect intersection turning movement and highway flow data to support the calculation of traffic signal timings. Making use of spreadsheets and other tools, the engineer can create a toolbox of data that is specifically designed to support the tasks involved in the job. Unfortunately, while this provides specialist support for the job in hand, it prevents an enterprise-wide view of data.

With the capabilities of data science today, it is possible to leave the cockpit intact while also copying the data to the data lake. Just like a real lake, it is then possible to use tools to waterski and to deep dive, exploring the data and revealing insights. It is worth noting that the concept of a data lake also implies that early judgment should not be applied regarding the usefulness of data. It is possible that a seemingly useless piece of data can combine with another piece of data in the data lake to create a valuable insight.

A fragmented data collection and management approach is analogous to a skilled worker such as a carpenter, who has assembled a collection of tools over the years and brings these tools to bear on each project. Carpenters do not look to a larger organization for these tools but expect to bring them as part of the experience and knowledge that they have gained over the years. This approach breaks down when specialist tools are required, and it may not be feasible to own these on a personal basis. It also does not provide the basis for sharing the cost of developing and implementing the tools.

The old English term bodger comes to mind. The term was first used to describe skilled carpenters who would set up an impromptu wood shop near a forest and, relying on their skills rather than specialist tools, create furniture and other wooden objects. These days the term refers to anyone who creates objects from a mishmash of found or improvised materials. While the skill required to be a bodger is laudable, the whole approach does not lend itself to the time savings and quality improvements that can be achieved using specialist tools on a shared basis. This is the essence of the conversion of data within an organization from silos and cockpits to an enterprise-wide data horizon. This also suggests that organizational and cultural change will be required to take full advantage of centralized data repositories, or data lakes.

The benefits of economy of scale, the ability to apply specialist tools to data management, and the conversion of data into information mitigates toward the centralization of data. In an ideal situation, centralized data would be used in combination with appropriate decentralized data, and the resulting information would be distributed in an optimal pattern across the enterprise. It would also be possible for anyone within the enterprise to view a catalog of available data and to be able to explore the value of enterprise data to the specific job function. In many situations, this is not the case, and data is kept in a computer or server next to a desk that is not visible to other members of the department or to other departments. The individual or team doesn't expect to rely on a central data repository or look to the enterprise to provide for information needs. While this is very efficient from the myopic viewpoint of the team or individual, it is inefficient in not supporting an enterprise- or organization-wide view of data. We have learned over the years that data is best used when it can be shared across multiple job functions. The old adage "put data in once, use many times" still holds good. Taking a function-specific view of data collection and management also prevents the achievement of cooperation and economy of scale. Fragmentation also causes duplication and hardware and software resources and provides a challenge with respect to configuration management—the need to keep one single version of the truth with respect to data for the organization.

Another issue with this conventional approach to data collection and management is that it becomes very difficult to know what data has been collected by the organization in its entirety. In the fragmented approach, it is highly likely that data is collected many times and perhaps used once, if at all. As a side note, experience in working with departments of transportation and cities in the United States indicates that a disproportionate value can be achieved through the creation of a data catalogue. While data analysts and data scientists would see this as a mere steppingstone toward the good stuff, it is obvious from a practical point of view that simply knowing what data the organization has collected and where it is located is extremely valuable to smart city and transportation professionals.

9.4 Definition of a Data Lake

Up to this point in the book, the focus has been describing big data and analytics and the questions to be addressed. This chapter focuses on the techniques required to build and manage a big data repository. One explanation of the term data lake is as follows:

> The idea of data lake is to have a single store of all data in the enterprise ranging from raw data (which implies [an] exact copy of source system data) to transformed data, which is used for various tasks, including reporting, visualization, analytics, and machine learning. The [data lake] includes structured data from relational databases (rows and columns), semistructured data (CSV, logs, XML, JSON), unstructured data (e-mails, documents, PDFs) and even binary data (images, audio, video) thus creating a centralized data store accommodating all forms of data [2].

The data repository is often referred to as a data lake, and this analogy will be used in this chapter. A data lake is a concept or analogy that is used to explain the centralization of data into a single repository. It is a collection of data from multiple sources that is accessible on an enterprise- or organization-wide basis and that takes advantage of the dramatically reduced cost of storing and manipulating data because of technologies such as Hadoop. It is a hardware and software environment that supports data sharing and supports the creation of a data catalogue. The creation of a data catalogue is an important dimension in the creation of a data lake as it informs the entire organization with respect to data available.

Data science capabilities continue to evolve and emerge. The latest evolution allows for the conduct of real-time analytics on data as it is being streamed from the collection point to the storage area. For the purposes of this book, this technique is referred to as the data river, as it involves processing on a stream, rather than a static body. It is feasible that real-time processing on data streams on the way to the data lake and analytics conducted on static data already in a data lake can be supported within one framework for managing data and information. A robust approach to the creation of a data lake would accommodate such data streaming analysis as well as the analysis of archived or static data.

9.5   How a Data Lake Works

Another perspective on data lakes that provides good insight into their nature and characteristics is to consider how they operate. Figure 9.2 shows a generic data lake configuration that can be considered as a model for a smart city data lake.

The major elements of the data lake are discussed in the following sections.

Data sources

One of the important aspects of a data lake lies in its ability to ingest data from multiple sources. Within smart city and transportation contexts, this would include infrastructure-based sensors such as traffic and passenger counters and probe data emanating from connected and autonomous vehicles. Data for the data lake could also be sourced from existing systems and databases such as those previously deployed for traveler information, traffic management, freight, and transit management. Data could also take the form of social media feeds such as twitter, image and video data from cameras, and other image processing–based sensors. A rich stream of data could also be sourced from smart phone apps operated by the public or private sector. In developing a roadway transportation data plan, the U.S. DOT [3] considers the following sources of data:

- *Infrastructure data:* Roadway geometry, roadway inventory, intersection characteristics, and the state of system controls.
- *Travel data:* Vehicle location, presence and speed within the system, internal vehicle status, transit vehicle location, speed and status, passenger counts, and schedule adherence data. Fred vehicle location and positioning with cross weight or data regarding the type and time critical nature of goods carried.
- *Climate data:* Prevailing weather and pavement surface conditions collected from roadway weather information systems (RWISs).
- *Modal data:* This includes border crossing data from U.S. customs and border protection regarding trucks, trains, containers, buses, personal vehicles, passengers, and pedestrians.
- *Travel behavior data:* Travel behavior, changes in travel characteristics over time, travel behavior related to demographics, and the relationship of demographics and travel over time

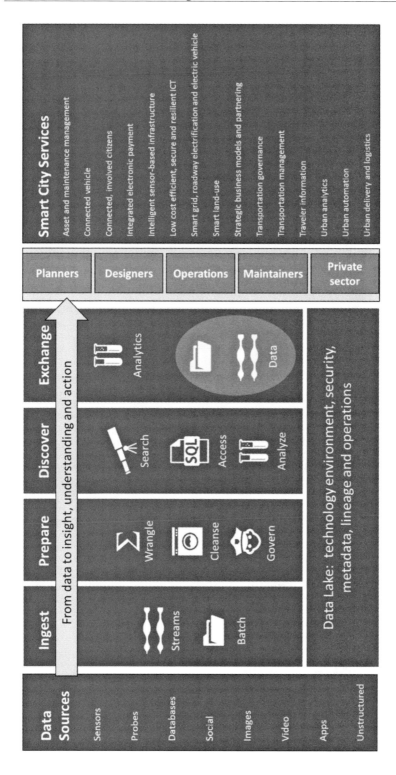

Figure 9.2 Model data lake configuration with major elements identified.

- *Other data programs:* Programs that provide access to federal data sets and tools across government agencies.

- *Real-time data capture and management:* The data resources testbed (DRT), the concepts and analysis testbed (CAT), and the cooperative vehicle highway testbed (CVHT).

Data coming into the data lake could also be unstructured data such as PDF files, e-mail, and other documents.

Data Ingestion

Data can take the form of static archived data or real-time streams from field devices and other sources. The Internet of Things will generate large volumes of data from sensors and other connected devices. The data is ingested into the data lake to create a single repository that can be accessed for data exchange and for analytics purposes. This activity would also include the establishment of suitable data-sharing agreements to enable the data to be accessed and shared in a manner to make it accessible to the data lake.

Data Preparation

The data preparation element consists of wrangling, cleansing, and defining governance arrangements for the data. Data wrangling can be a manual or semi-automated process making use of decision support tools to bring data to common formats and locational referencing systems. Data would also be verified at this stage by comparing the same data from different sources and identifying potential gaps or weaknesses in the data. Duplication and errors are removed in a cleansing process as part of data preparation. At this stage, data governance arrangements are identified to manage data sourcing, data access, and data distribution. This will also include arrangements for sharing analytics that are derived from the data lake during the data discovery process. The U.S. DOT in developing a roadway transportation data business plan [3] also noted the need to address data quality. U.S. DOT recommendations include the development of a policy to define responsibilities for data quality and adopting data quality standards for data collection, processing, application, and reporting.

Data Discovery

In the data discovery element, data is searched, accessed, and analyzed. A range of search and access tools such as structured query language (SQL) and other statistical functions can be used to detect trends and patterns in the data to reveal new insights and understanding. Analytics functions that could be used during data discovery include statistical, cluster analysis, data transformation, past, pattern and time series, decision tree, text, and graphic.

Data and Analytics Exchange

This element of the data lake supports both data exchange and analytic sharing. Experience shows that providing multiple users with data is not enough to motivate and enlighten them regarding the use of the data for their own purposes. It is also important to share analytics for two major reasons. In the first case, analytics may be directly applicable to the user's job function. In the second case, the analytics can be used as a communication tool or model to illustrate how analytics could be applied to the user's job function.

The ingestion, preparation, discovery, and exchange of data is supported within the data lake environment that includes both hardware and software. In addition to supporting these essential functions, the data lake environment can also support the appropriate security arrangements and the definition of metadata, data lineage and other master data operations required to maintain a single version of the truth and catalog the data within the data lake. The data lake technology environment would also support operations such as the administration required to keep the data lake running.

Delivery of Insight and Understanding to Smart City Practitioners

The analytics that result from the discovery process are presented to smart city practitioners to form the basis for response strategies in the light of the insight and understanding. The whole data lake configuration is designed to support a strong connection between data and such insight and understanding. To harness the full value of the analytics it is necessary to develop actionable work items and strategies that represent a response to new information regarding prevailing transportation conditions and the quality of transportation service delivery in the smart city.

Support for Smart City Services

Ultimately the strategies derived from new insights and understandings will provide support for the 16 smart city services defined in Chapter 5. For example, with respect to the smart grid, roadway electrification, and electric vehicle, the output from the data lake could provide insight into optimum placement for the electric vehicle charging points. It could also provide insight into energy requirement changes that would result from large-scale adoption of electric vehicles in the smart city. With respect to the integrated electronic payment service, the data lake could provide insight into the optimum fee or ticketing structure to maximize user experience, minimize operating costs, and ensure that revenue is as predicted.

It is expected that support for smart city services would cover the spectrum of transportation activities from planning, design, and operations through maintenance. The support would address both public- and private-sector needs.

This indicates that the private sector may also have access to both data and analytics from the data lake.

9.6   Value of a Data Lake

While the ultimate value of a data lake lies in providing support for smart city services and the delivery of insight to smart city practitioners, there is a range of benefits that are realized when a data lake is developed. These include the following:

- Enterprise-wide data access for timely analytics and insights;
- The foundation for large-scale proactive analytics;
- A steppingstone toward automation through predictive analytics and machine learning;
- Reduced costs due to data management duplication and processing duplication;
- Improved safety, efficiency, and user experience by accelerating analytics work;
- Better data governance with a single consistent version of the truth and better control on who, what, and when data is accessed or provisioned;
- The ability of data elements to combine for new analytics;
- Discovering value in unused data and relationships between data sets regarding customer behavior and transportation service delivery quality;
- Providing a platform for innovation in smart cities and transportation;
- Providing support for smart city service delivery.

Now we consider each of these in turn.

Enterprise-wide Data Access for Timely Analytics and Insights

While bringing the data together in creating the data lake, the data assets for the smart city become visible. In addition to creating a data catalog that enables all users to see the data that has been collected, the data is made available to a wide range of users for further analysis. The analytics that have been derived from processing data in the data lake can also be shared across the organization, providing motivation and stimulus for further use of the data lake and development of customized analytics for specific job functions such as transportation planning, traffic engineering, and asset management.

The Foundation for Large-Scale Proactive Analytics

The data lake is also the basis for further efforts related to large-scale proactive analytics. While this will also require cultural and organizational change, the existence of the data lake opens the way for the smart city organization to apply large-scale analytics that will guide many aspects of planning and delivery for smart city transportation. This enables the adoption of results-driven actions and the establishment of scientific approaches to transportation service delivery, based on observation, understanding of mechanisms, and data.

Steppingstone Toward Automation Through Predictive Analytics and Machine Learning

There is considerable interest in activity in the concept of an automated vehicle, and it would seem relevant to also consider how automation can be applied to back-office processes in the smart city. While it may not be appropriate or even desirable to leap toward an automated back office overnight, the establishment of analytics and the development of ability to make predictions can form the basis for the past toward automation. The availability of the data in the data lake can also form the raw material for the support of machine learning and deep learning techniques that support the stated development of artificial intelligence in the smart city back office. It is likely that this will begin with sophisticated decision support for the humans involved, with full automation a possibility over the longer term.

Reduce Costs Due to Data Management Duplication and Processing Duplication

Adopting a fragmented approach to data collection storage and management will inevitably lead to duplication. In fact, the cost of duplication may be buried within the overall cost of operating and maintaining the current data collection, storage, and management system. The process of creating a data lake is likely to shine a light on the volume of duplication and provide estimates of the costs involved. Cost savings are likely to be identified in data collection, as well as data storage and processing. Based on experience, the average transportation agency supports multiple redundancy with respect to data collection, with considerable amount of ad hoc data collection for project-specific purposes. If such data is not visible across the organization, then it is likely that other ad hoc initiatives will collect the same or similar data. In some cases, awareness of the data is insufficient, and an inability to access the data in a reasonable time frame forces project specific data collection to go ahead even if duplication is understood. Cost savings are also likely to be realized with respect to software licenses. Multiple software licenses may have been procured to support a fragmented approach to data storage and management. As the data lake is created, opportunities may be revealed to save money by consolidating software licenses.

Improved Safety, Efficiency, and User Experience by Accelerating Analytics Work

There is often a significant time lag between smart city practitioners understanding the need for data analytics work and the work that has been put into practice. The creation of the data lake and the number of self-service tools and opportunities present their ability to parallel-stream work efforts and shorten the time from realization of need to satisfaction of need with respect to data analytics.

Better Data Governance with a Single Consistent Version of the Truth, Better Control on Who, What and When Data Is Accessed or Provisioned

Bringing data together in a data lake allows better possibilities for data governance and configuration control. A single repository is much easier to manage with respect to access control on data going into the lake and data analytics emanating from the lake. A single repository also makes it easier to upgrade software at a single point rather than at multiple points in a fragmented system.

Enabling Data Elements to Combine for New Analytics

The new big data approach to data storage and manipulation allows us to delay judgment on the value of data. Data storage costs have reduced to the point where a new strategy can be implemented. In simple terms this strategy involves the capture of as much data as possible; the data will be ingested into the data lake and then allowed to demonstrate its value. In this scenario, a seemingly useless piece of data may combine with other data in the data lake to create a new and valuable insight. There is a significant element of discovery that is supported by the establishment of a data lake that would be unachievable in a fragmented data storage and management approach.

Discovering Value in Unused Data and Relationships Between Data Sets Regarding Customer Behavior and Transportation Service Delivery Quality

Building on the previous point, new value can be realized from data that has either been hidden or unused within the overall smart city organization. Sunk investment in data collection can be revitalized through the discovery of new uses for the data. This also extends to the understanding of new relationships between data sets. This is particularly relevant with respect to customer behavior and monitoring of transportation service delivery. Deeper insights into how travelers behave and the prevailing transportation conditions they face will ultimately lead to better strategies and tactics in the smart city.

Providing a Platform for Innovation in Smart Cities and Transportation

The use of analytics and big data techniques is only a small part of the overall innovation that can be achieved by a smart city. However, the focused nature of the work in creating a data lake can be used to propel the use of big data and analytics; it can also form a stimulus for innovation on a wider front within a smart city. Deeper insights into and understanding of travel behavior and transportation service delivery can provide the raw material on which to design innovative strategies and services for both the public and private sector.

Providing Support for Smart City Service Delivery

Finally, the establishment of a data lake will provide support for smart city service delivery. By enabling the management and measurement of smart city service delivery, the data lake provides better management possibilities and richer opportunities for smart city service delivery improvement.

9.7 Challenges

Based on prior experience, there are many challenges that must be addressed when building a data lake. These can be summarized as follows:

- Lack of clear strategy;
- Existing data scattered and not well understood;
- Difficulty in turning data into action;
- Lack of big data skills;
- Insufficient governance and security;
- Degradation of the data over time without data quality control;
- Lack of self-service capabilities and long development times;
- Lack of features to motivate and enable smart city and transportation exponents.

Each of these challenges is explained in the following sections.

Lack of Clear Strategy

Like many information technology tools, the hardware and software environment required to create a data lake can be acquired with little thought to the overall strategy or endgame. While this enables rapid progress, if the direction and the ultimate destination are not known or ambiguous, it is highly possible that the entire initiative will end up in a dead end. It is necessary to have a clear strategy that incorporates the goals of the exercise and clearly articulates the value and benefits. From a smart city and transportation perspective, it is important to define the safety, efficiency, and user experience objectives and to define how the proposed data lake will fit in to the overall picture of data and information exchange on a citywide basis.

Existing Data Scattered and Not Well Understood

It is highly likely that the existing transportation data is scattered and not well understood. There may or may not be an existing data catalogue, and even if one exists, it may not be complete and up-to-date. To create a data lake, it is necessary to identify the sources of data and plan to have access to the data that will be placed in the data lake. In many cases this can take a considerable amount of time and consume significant resources.

Difficulty in Turning Data into Action

Bringing the data together into a data lake does not guarantee results. To harness the value of the data lake, it is necessary to support an entire process that results in actionable insights and the development of strategies to be applied in response to the new insight and understanding. In many cases, this may require some organizational adjustment to empower staff to take advantage of the analytics developed from the data lake.

Lack of Big Data Skills

The use of big data techniques and analytics is relatively new to transportation in smart cities. Therefore, it is likely that the big data skills required to successfully establish and operate the data lake may not exist within smart city or transportation organizations. When planning for the establishment of a data lake, it will be necessary to identify the required skills and decide how those skills will be sourced, whether by outsourcing or new hires.

Insufficient Governance and Security

The adoption of a bottom-up approach that is not guided by a clear strategy or an unambiguous understanding of the final big picture can lead to insufficient governance and security. Taking advantage of the power and flexibility of available technology can support rapid progress, but it can also allow essential activities related to governance and security to be bypassed.

The Degradation of Data Over Time without Data Quality Control

An unfortunate trend in the application of advanced technologies to transportation is the creation of a trajectory for technology application. In the trajectory, considerable progress is made in the implementation of advanced technologies, and the target levels of service are attained. These service levels then degrade over time as insufficient resources are allocated to operations and maintenance of the initial technology deployment. The same challenge exists with respect to data lakes for smart cities and transportation. It is necessary to take steps to not just launch the data lake, but to ensure that it does not degrade over time. This also includes taking the necessary steps to monitor, manage, and ensure data quality both entering the data lake and being maintained within the data lake.

Lack of Self-Service Capabilities and Long Development Times

Very often in technology-driven data lake implementations, the data lake is developed as a tool that requires highly specialist operation. This leads to a situation in which end users cannot have direct access to the data or the analytics. The absence of such self-service capabilities can produce a heavy workload on a few members of staff, leading to long development times and slow responses to end user needs.

Lack of Features to Motivate and Enable Smart City and Transportation Exponents

In a similar vein to the above challenge, an information technology–driven data lake program can ignore the need to motivate and enable end users in smart city and transportation contexts. This can lead to a lack of interest on the part of the end users and a consequent inability to monetize the investment made in the data lake. Early experience indicates that it is not sufficient to merely share data; it is also necessary to provide models and illustrations that can motivate the end users. This would include helping end users to understand the data lake and to understand the analytics possibilities through the communication of model analytics and the support of a dialogue on the development of custom analytics for the end user's specific job function.

9.8 An Approach to Building a Data Lake

Early experience in the creation of data lakes for multiple organization types in both the public and private sector has revealed that there are a few pitfalls to be avoided in the successful development of a data lake. Learning lessons from this early experience, it is possible to put together a robust approach that minimizes the chances of encountering these barriers while maximizing the chances of success. To provide practical advice on the creation of a transportation data lake, a data lake creation methodology has been identified and is defined in subsequent sections of this chapter. The approach is based on the experiences of a company called Think Big [4], and it has been evolved as a direct result of experience gained in working with public- and private-sector clients. The original approach methodology has been adapted based on experience with several transportation clients to create an approach that is specifically adapted to the needs of transportation and smart cities. Figure 9.3 presents an overview of the approach methodology.

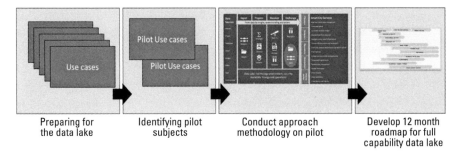

| Preparing for the data lake | Identifying pilot subjects | Conduct approach methodology on pilot | Develop 12 month roadmap for full capability data lake |

Figure 9.3 A proposed approach methodology for the implementation of a smart city and transportation data lake.

The overall philosophy behind the approach is to make incremental or stepwise progress toward the establishment and operation of the data lake. Initially the focus is placed on a very small number of use cases that are addressed in a pilot project. The pilot use cases are then used as the basis for conducting the approach methodology on a pilot project. During the pilot project, all elements of the operation of the data lake are brought into play, including ingestion, preparation, discovery, and exchange of data. Each step in the methodology is explained in the following sections.

Preparing for the Data Lake

Preparation or planning for the data lake includes exploring requirements and objectives with the target end users. In the case of a smart city transportation initiative, end users typically consist of city officials and other city transportation partners such as departments of transportation, transit agencies and other transportation service providers. It would also be helpful to bring relevant private-sector participants into the dialogue at this early stage.

Identifying Pilot Subjects

The use case concept is employed as a major tool in identifying subjects for the pilot. The use case pilot described in Chapter 5 can be an effective format for capturing the use case information. Whether the format is used or not, it is essential to capture the primary ingredients for the pilot, including the data to be used, the expected value to be achieved, and an initial understanding of the analytics to be applied. The selection criteria for the use cases are as follows:

- The use cases deliver immediate value to the city.
- Data to support the analytics required for the use cases is readily available either from city sources, transportation partners, or the private sector.

The overall intent is to deliver early results that enable the communication of the potential for the data lake. The early results also enable an effective dialogue with our range of users who might find the data and analytics useful; this also promotes interest and activity around the use of the data lake for multiple purposes.

The delivery of business value to smart city exponents is also crucial in the implementation of the pilot project. The pilot implementation allows the focus to be moved from data and data management to the delivery of insight and understanding and the subsequent incorporation of these into new strategies and new ways of doing business. The definition of business value involves the identification of the objectives to be addressed by the analytics and the problem statement that summarizes the need issue problem to be addressed. At this stage in the process, a catalogue of available data sources is created. This represents the state of available data and captures the data's format, structure, and current location. This would also include the establishment of suitable data-sharing agreements between smart city transportation partners. It is also beneficial to develop a preliminary list of analytics that will be conducted on the pilot data lake along with an identification of the proposed users of the analytics and the likely benefits that will be attained through the availability of the new insights and understanding. This will form part of an overall summary.

Conducting the Approach Methodology on the Pilot Project

The entire approach methodology is conducted on a pilot basis, focusing on the use cases selected for the pilot. This allows experience to be gained in data ingestion, preparation, discovery, and exchange. During this stage of the pilot, data governance and data exchange arrangements would be defined and put into pilot operation. This would also include significant dialogue with end users on the use of the pilot use case analytics, on the possibility to extend the data lake. The conduct of the analytics should address other job-specific needs.

Developing a 12-Month Roadmap for a Full-Capability Data Lake

Based on the results of the pilot and the experiences gained, a 12-month roadmap for a full-capability data lake can be prepared. While the exact contents of the roadmap will depend on the needs of the city and organizations in question, it would typically contain the following at a minimum:

- A full set of use cases to be addressed and supported by the full-capability data lake;
- Sequencing of the use cases across the 12 months;
- The development of a six-month action plan with required investment and business justification based on the experiences of the pilot project;

- The development of plans and briefing materials for senior executives to form the basis for further decision-making;
- The definition of a complete hardware and software environment or architecture to support the full-capability data lake;
- An implementation plan including activities, schedule, and budget estimates for the expansion of the pilot project to full capability.

Table 9.1 provides a summary of how the proposed approach meets the challenges defined in Section 9.7.

9.9   Organizing for Success

It is hoped that when a transportation organization embarks on the creation of a data lake, the effort will also result in a new approach to data collection and acquisition. All too often in transportation, data is collected on a speculative basis with little concern for the eventual use of such data. The establishment of a data lake and organizational alignment to the data lake should provide some guidance and insight into a new approach to data collection based on the needs of the data and the need for which the data has been collected. Ideally, data collection and acquisition will lead to the conversion of data to information using analytics, and experience gained using analytics will provide feedback on the need for additional or higher-quality data. In this respect, the data lake cells function as a feedback mechanism to guide data collection and acquisition. In an ideal environment, data collection and acquisition will be driven by a clearer understanding of the use to which the data will be put. For example, the use of the initial data lake to conduct some preliminary analytics may result in a much more detailed understanding of the completed data set required to get results. This will also provide insight into the required accuracy of the data. While early results are not invalid as they may provide insight that was not previously available, even better results may be possible with better data. Accordingly, the use of a data lake can provide significant input into a structured and planned approach to data collection and acquisition. Prior experience in the application of information and communication technologies to transportation within a system engineering framework has indicated that, in most cases, the technology solution that results from the planning and design process represents a theoretical ideal. Approaches to the development of system architectures assume that the ideal technology solution will be applied and that it will be necessary to adjust or fine-tune organizational arrangements to match the needs of the technology solution. It is more likely that the ideal technological solution will be adjusted and perhaps suboptimized to fit existing organizational arrange-

Table 9.1
How the Approach Addresses the Challenges

Data Lake Challenges	How the Approach Addresses the Challenges
Lack of clear strategy	The approach allows the development of a clear strategy that can be tested and practiced using a pilot. This strategy will then be revised and enhanced in the light of lessons learned and information received during the pilot.
Existing data scattered and not well understood	The adoption of an incremental approach allows early results while providing the time necessary to bring scattered data together. Early results will also act as a communication tool to motivate staff to identify additional data and help to bring it together.
Difficulty in turning data into action	The delivery of early results illustrates the full process from data collection to creation of a data lake, demonstration of ability to turn data into action, and the development of real-life strategies to serve as models for future analytics.
Lack of big data skills	The approach enables both public- and private-sector resources to be combined and transitions to be created from initial project to full-scale project.
Insufficient governance and security	Data governance and security arrangements are tested during the pilot project.
After initial establishment, degradation over time without data quality control	Data quality control measures can also be tested and developed during the pilot project, forming a practical platform for preventing degradation of operation over time.
Lack of self-service capabilities and long development times	The ability to share data and analytics in pilot use cases enables the establishment of self-service capabilities and shortened development times using agile development approaches.
Lack of features to motivate and enable smart city and transportation exponents	Early delivery of practical results provides the tools necessary to motivate smart city and transportation exponents on the use of the data lake. This also stimulates exponents to consider their own use of the data and what further analytics would be required. This has the added benefit of enabling results-driven data collection and acquisition.

ments. This is unlikely to provide the best return on investment for the creation of a smart city data lake. In order to realize the new potential and possibilities offered by the data lake, it is necessary to implement organizational change and to build awareness among end users regarding the potential. One possible way to address this challenge is the development of an organizational plan as part of a pilot project. The transportation data analytics that emanate from the pilot project can also be used to support pilot arrangements for fine-tuning the organization. This may also require a cultural change that focuses on the organization's ability to adopt innovative techniques rather than following existing processes and procedures. In this respect, data analytics can be used as a bridge from the data lake to job functions for end user staff. I have grappled with the issue of how to develop organizational arrangements that would be the best fit for both technological and commercial layers of the architecture for some time.

I have decided that perhaps the solution lies in the application of transportation data analytics. Organizational change would take the form of changes in job objectives and descriptions and the adoption of a new culture. To address the idea of a new culture, consider the following football analogy.

World-class reporting will enable staff within an organization to be extremely well-informed spectators at the football game. Great transportation analytics will empower the same staff to be coaches and exert influence on the performance of the organization, our team. It is expected that transportation data analytics will significantly impact the planning, design, and operations of transportation while providing guidance for future transportation investment programs. As an example, an interesting analytic might be dollars per percentage modal shift toward public transportation. This would be a measure of the effectiveness of investments designed to influence the modal shift in the region in favor of public transit.

So how exactly would transportation data analytics shape the organizational arrangements within a transportation enterprise? Analysis of big data sets using suitable discovery tools will reveal the trends, patterns, and underlying mechanisms of transportation. Transportation data analytics will be defined and can be used to manage planning, design, implementation, operations, and maintenance of transportation. These transportation data analytics would then be incorporated into job descriptions and objectives for the staff involved in transportation service delivery.

A simple example would be the role of the traffic signal engineer. At the moment, the job of the traffic signal engineer is to run the traffic signal system. Perhaps in the future, the job objectives of the traffic signal engineer will be stated as the minimization of stops and delays across the corridor or the network. There may also be other advanced analytics yet to be discovered that could be suitable for use as job objectives. The job description for a traffic signal engineer would then be written around the attainment of the job objectives.

For example, a narrow view of this would suggest that the change in emphasis could be unfair to the traffic signal engineer, as some factors affecting stops and delays are not within his or her control. Perhaps the job description could be written to include the need to cooperate and collaborate with others whose actions affect the primary objectives. This also leads to the thought that perhaps ITS user services could have analytics associated with them for the purposes of measuring the effectiveness of the delivery of the services. This concept holds out the possibility of building a bridge between the various layers of an ITS architecture, while also setting the scene for a laser focus on results rather than activity.

The emergence of big data and the connected vehicle and growing understanding of the data science possibilities from outside of transportation means that our business is going to change. Perhaps this is the ideal opportunity to take a close look at how we organize for success?

9.10 Summary

The creation of a data lake requires the application of hardware and software and suitable organizational change. To attain the full benefits of the data lake, it is necessary to use the data, the analytics, and the new insight and understanding that is made available to develop responses, strategies, and actions that will support the delivery of smart city transportation services. This chapter provides a detailed definition of a data lake, along with an exposition of the various elements that are brought into play when a data lake operates. The chapter aims to explain the term data lake and to show, at a planning level, how it can be implemented. However, it does not provide guidance on the selection of specific technologies, as there are many such options. This is beyond the scope of this book.

The chapter also provides an overview regarding the challenges that are likely to be encountered in the creation of a data lake for a smart city, along with a summary of the likely benefits that can be achieved by adopting the data lake strategy. Advances in data science have enabled us to aggregate data in ways that were not possible in the past, allowing data to combine in new and interesting ways, while providing an enterprise- or organization-wide horizon on the data.

The innovative nature of the data lake also presents an opportunity to reevaluate the shape and structure of smart city and transportation organizational arrangements, with respect to data analysis. Consequently, the chapter concludes with some initial thoughts on how organizational fine-tuning might be achieved using the data lake and data analytics as an important tool in the process. It is difficult to overemphasize the importance of the pilot project approach and the methodology. Many of the technologies and the resulting insights will be alien to smart city and transportation professionals, and the conduct of a pilot provides an opportunity for awareness and understanding that will support the extraction of the best possible value from the data lake investment. The approach also supports a focus on actionable insights and enables smart city and transportation staff to maintain their focus on transportation service delivery and results.

References

[1]   Hadoop definition, tech target.com http://searchcloudcomputing.techtarget.com/definition/Hadoop, retrieved January 17, 2017.

[2]   Data Lake characteristics from Wikipedia, https://en.wikipedia.org/wiki/Data_lake#cite_note-stein2014-4, retrieved January 18, 2017.

[3]   U.S. DOT roadway transportation data business plan (phase 1) – final, https://ntl.bts.gov/lib/48000/48500/48531/6E33210B.pdf, retrieved April 10, 2017.

[4]   Think Big Data Lake Foundation, https://www.thinkbiganalytics.com/data-lake-foundation/, retrieved on January 17, 2017.

10

Practical Applications and Concepts for Transportation Data Analytics

10.1   Learning Objectives

This chapter addresses the following learning objectives:

- It provides practical examples of how analytics can be applied to transportation subjects;
- It explains how analytics can be used for speed variability in freeways;
- It describes how analytics can be used to measure accessibility in a smart city;
- It explains the application of analytics to develop a performance analytic for toll roads;
- It defines an analytic approach for arterial performance management;
- It explains how analytics can be used to support better decision-making for transit bus acquisition;
- It illustrates the use of analytic techniques;
- It discusses analytics and how they can be applied to other subjects.

Figure 10.1   Word cloud for Chapter 10.

10.2   Chapter Word Cloud

A word cloud, shown in Figure 10.1, has been prepared for this chapter to provide an overview of content.

10.3   Introduction

This chapter describes implementations and concepts for the application of analytics to transportation. One of the challenges in explaining big data and analytics in transportation is to show a strong connection between user needs or the real situation to be addressed, with the capabilities of data science and analytics. While it is likely that data science and analytics can address almost every transportation problem, experience has shown that most progress is made in the application of data science to transportation when a narrower focus is placed on specific areas of need. To narrow down the focus to the practical application of big data and analytics techniques, five concepts have been identified: freeway speed variability analysis, smart city accessibility analysis, toll return index for toll road performance, arterial performance management, and decision support for bus acquisition. The concept of freeway speed variability, which was implemented in cooperation with a client, has been the subject of extensive development and application. In the other cases, the concepts have been developed in coordination with a range of potential users but have not yet been implemented. In any event, all the concepts shed significant light on how data science can be applied to practical needs within the smart city and transportation realms.

10.4 Concepts

The concepts discussed in the following sections were defined to illustrate the application of data analytics techniques to transportation.

Freeway Speed Variability Analysis

This concept entails the application of big data and analytics techniques to understanding speed variability in bottleneck formations on a major freeway. This concept was applied to a major urban freeway in close cooperation with a large Department of Transportation in the United States. The experience provided some rich insight into the use of analytics for traffic engineering under freeway conditions and yielded some lessons regarding the data quality and the analytic discovery process. This concept has the largest amount of documentation in this chapter, because of this practical experience.

Smart City Accessibility

This concept entails adopting an analytics approach to defining the ease or difficulty of movement from residential zones to work, healthcare, and educational opportunity zones. It was developed in close cooperation with a city in the United States, implementing a smart city initiative. One of the core objectives of the smart city initiative is to improve accessibility to jobs, healthcare, and education opportunities. The concept is at an advanced stage of development but has not yet been implemented.

Toll Return Index

This involves the creation of a composite toll return index that communicates the benefits delivered by a toll agency in return for the toll paid. The toll paid is divided by the total benefits delivered to create the index. The benefits are comprised of three primary components: safety, efficiency, and user experience. The concept has been discussed and evaluated by both current and former executive-level toll management, but it is still in the course of being implemented by toll agencies.

Arterial Performance Management

This concept involves the application of analytics techniques related to traffic turbulence to determine the technical performance of arterial corridors and networks. It is supplemented by social sentiment analysis and context-specific keywords analysis of Twitter feeds to determine driver perception in addition to technical performance of the corridor.

Decision Support for Bus Acquisition

This concept entails the application of analytics to develop a decision-support tool for bus acquisition. The concept considers the prevailing traffic conditions encountered by buses as they service their bus routes and provides decision support to identify the optimum timing for the acquisition of new buses. The concept was developed in cooperation with a large suburban bus operator, but it has not yet been put into practice.

Each of these concepts is described in more detail in the following sections.

10.5   Freeway Speed Variability Analysis

The analysis of speed variability on freeways is an important subject that can yield significant insight into the operational efficiency of the facility. This following section explains how analytics can be applied to this important subject.

Overview

Not every highway has the appropriate sensors installed to allow traffic speeds to be determined. However, several private companies have invested in the ability to collect probe data from both vehicle fleets and individuals using the appropriate smart phone app. With the application of suitable software, it is possible to take fleet data and individual data as a sample, extrapolate it, and combine it to provide a comprehensive speed data set across a city. While the data is based on a sample, it allows for the creation of a comprehensive data set for an entire citywide area. Making use of this data source, an evaluation was conducted on the effects of variable speed limit (VSLs) on a major freeway in the United States.

The problem addressed relates to understand the effects of applying VSLs to a major urban freeway. Conventional approaches may support the level of detail analysis required to identify specific effects. The objective of the VSL implementation was to address dramatic speed reductions within slowdowns or bottlenecks, caused by recurring or nonrecurring congestion. The work was implemented as a proof of concept in cooperation with the U.S. Department of Transportation. Over the course of discussions with the Department of Transportation, interest was identified in the use of big data and data analytics for transportation. It was also identified that the data science and data analytics world represents a separate community from the world of the transportation professional. To create a bridge between these two specialized and very important communities, it was decided that a proof-of-concept exercise would be conducted centered on an evaluation of the state DOT's VSL project.

VSL is a concept that has been adopted in many parts of the world but is relatively new to the United States. The quantification of customer service, safety, and efficiency benefits of VSLs is extremely important to transportation practitioners. This fits within a wider need to understand the specific effects of investments in smart city services and other transportation systems and infrastructure. While there is a great deal that can be done with data science and analytics in transportation, this approach decided to focus on this relatively narrow application as it provides an ideal opportunity to demonstrate the range of capabilities of data analytics and provides direct input to a subject area that is of importance to the state department of transportation in question.

The VSL approach involves the application of dynamic message sign technology to the control of traffic on interstate highways (freeways) around a major urban area. Dynamic message signs placed at regular intervals along the freeways and connected via fiber-optic communications to a traffic management center are used to display mandatory speed limits to drivers. The prevailing speed limits are determined by measuring traffic conditions along the section of freeway using roadside sensors.

As explained on the state DOT website:

> VSL are speed limits that change based on road, traffic, and weather conditions. Electronic signs slow down traffic ahead of congestion or bad weather to smooth out flow, diminish stop-and-go conditions and reduce crashes. This low-cost, cutting edge technology alerts drivers in real time to speed changes due to conditions down the road. More consistent speeds improve safety by helping to prevent rear-end and lane-changing collisions due to sudden stops. Our ability to remotely change the speed limit on the corridor is not intended to create speed traps. Rather, the changing speed limits are designed to create safer travel by preventing accidents and stop-and-go conditions.

The section of the regional road network that was the subject to this proof of concept was equipped with 88 electronic speed limit signs at locations approximately a half mile to one and a half miles apart. The signs were located on the outside shoulder and the median of the highway. The ability to change the speed limits remotely is designed to create safer travel by managing stop-and-go traffic conditions by creating smoother traffic. Speed limits are adjusted in 10-mph increments from 65 miles per hour to a minimum 35 mph.

The VSL project began operation in September 2014, with the following stated objectives.

- Reduce congestion and traffic delays by harmonizing traffic flow and reducing traffic crashes;
- Reduce travel times;

- Reduce crash frequency, severity, and the likelihood of secondary crashes by reducing the speed of vehicles as they approach an incident, traffic queue, or stoppage;
- Stabilize and smooth traffic flows (consistent speeds within lanes and between lanes).

VSL has been implemented in other cities and studies show the following.

- VSL allows travel at a slower, but more consistent speed, as opposed to the constant stop-and-go traffic typical of rush hour conditions.
- By regulating traffic speed, VSL also helps reduce rear-end and lane-change collisions associated with sudden stops at the back of congested areas.
- This more consistent speed improves safety, saves motorists gas, and lessens harmful emissions from idling in stopped traffic.

Approach

The overall approach to the work involves a before-and-after analysis of traffic speeds at one-minute increments on segments that comprise the study area. To understand the underlying patterns of traffic variation along freeways in the study area, the before data set consisted of almost two years, or seven quarters of prior data from September 2012 to September 2014, and one quarter of post data (quarter four, 2014). The VSL project went into operation in September 2014. This limited the scope of the after data set to a single quarter, quarter four, 2014. It was considered that three months of after data would be sufficient for the effect of the implementation to stabilize. The following objectives were defined for the evaluation:

- To demonstrate the power of data analytics on an application that is within the state DOT's current focus;
- To illustrate the application of external professional resources for data analytics;
- To demonstrate how the application of analytics and predictive technologies can optimize the use of internal staff resources.

A review of previous evaluations of the VSL project suggests that there are two primary approaches to the evaluation of VSLs. The first includes the use of a traffic simulation model, and the second involves the analysis of before-and-after data. Since the VSL project has already been deployed and one of the key objectives of the work was to demonstrate the use of analytics on a large observed data set, it was decided to adopt the latter approach.

Initial Analytics Approach—Speed Variability

The analysis makes use of privately sourced Inrix speed data collected from 2013 to 2014 on the subject freeway, approximately 36 miles in each direction. There are 138 TMC locations for which Inrix [1] data was available on a minute-by-minute basis.

While direction signs on the freeway make use of formal directional designations based on compass headings at various locations of the freeway (east/west or north/south), for purposes of the analysis, the data was arrayed in clockwise and counterclockwise directions, proceeding clockwise from mile marker 9.97 in the west to mile marker 46.3 in the east, and vice versa proceeding counterclockwise from mile marker 46.5 in the east to mile marker 10.0 in the west. Note that the data analytics were conducted in Teradata Aster, with summaries and visualizations developed in Tableau.

Figure 10.2 shows a summary of the available data for the two directions for the full year and illustrates the speeds when considering only weekdays and then only weekday peak periods. An analysis of the data was conducted to

One minute speed readings by TMC for 2013 & 2014

Year of Thed..	Loop		Grand Total
	Clockwise	CounterClockwise	
2013	34,318,350	33,278,400	67,596,750
2014	33,404,175	32,673,104	66,077,279
Grand Total	67,722,525	65,951,504	133,674,029

Weekdays

Year of Thed..	Loop		Grand Total
	Clockwise	CounterClockwise	
2013	24,473,592	23,731,968	48,205,560
2014	23,865,636	23,345,551	47,211,187
Grand Total	48,339,228	47,077,519	95,416,747

Peak Periods

Year of Thed..	Loop		Grand Total
	Clockwise	CounterClockwise	
2013	9,186,474	8,908,096	18,094,570
2014	8,981,320	8,785,608	17,766,928
Grand Total	18,167,794	17,693,704	35,861,498

Figure 10.2   Summary of available data for two directions for the full year.

determine that the morning peak occurs from 6:00 a.m. to 9:00 a.m. and in the evening from 3:00 p.m. to 7:00 p.m. Using this definition, the peak periods span about 27% of the available data points.

Figure 10.3 shows the data for the fourth quarter, as the fourth quarter was the only quarter for which a direct comparison could be made between quarter four, 2013, without VSLs and quarter four, 2014, with the VSLs implemented. The VSL system was implemented in September 2014.

Figure 10.3 illustrates the speed variability for the entire study on weekdays during peak periods. This encompasses eight separate quarters, or 520 weekdays of data. The shading indicates the average speeds, and the width of the line represents again the standard deviation of the traffic speeds. Figure 10.3 illustrates that traffic conditions are in fact highly variable on the freeway, even during some of the most heavily traveled time periods.

Figure 10.4 shows the speed profile that was assumed for a bottleneck. This represents a formal data definition for a subjective traffic engineering term: bottleneck.

Figure 10.4 is adapted from industry best practice with respect to the definition of a bottleneck in terms of speed [2]. The bottleneck begins when the observed speed of traffic falls below 60% of the reference speed. The reference speed is an INRIX [1] data parameter, which represents the 85th percentile of the observed speeds. This roughly equates to the speed limit enforced on fixed speed limit signs along freeways. The slowdown is then assumed to persist until the traffic begins to climb back through 60% of reference speed. Note that this simplified template describing a bottleneck assumes that there will be only one slowdown per bottleneck and that when the traffic climbs back through 60% of reference speed, the bottleneck is over. There is the potential to define a chain of bottlenecks with one reduction in speed, when, in reality, the chain of bottlenecks could be one single bottleneck. This will be discussed later in this section.

Figure 10.4 illustrates speed variability across a length of freeway that is split into a series of TMC sections. A TMC segment is an industry standard definition of a section of freeway. Note that this should not be confused with the more general meaning of TMC as a traffic management center. This definition goes back to an earlier European project known as the Radio Data System—Traffic Message Channel project (RDS—TMC). Segment definitions were developed especially for this project to enable location-specific traffic messages, and they have been adopted on a wider basis since the project was completed. INRIX [1] adopted TMC segments as the basis for the current version of the data that was used for this work.

The vertical axis of the graph in Figure 10.4 represents traffic speed in miles per hour, averaged over one-minute time increments. The horizontal axis represents distance, with the arrow indicating the direction of travel for the traffic.

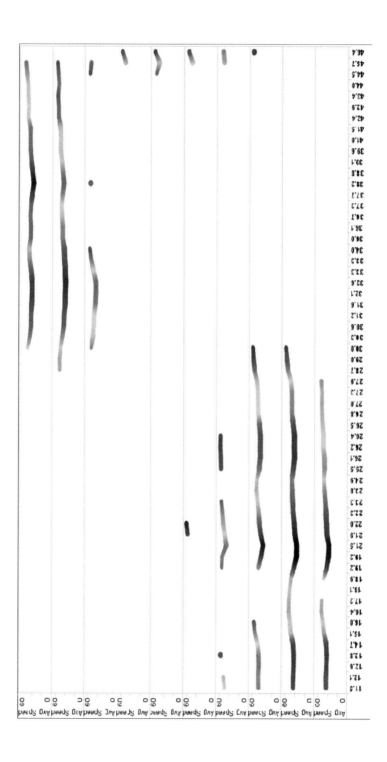

Figure 10.3   Speed variability for 2013–2014 peak period weekdays.

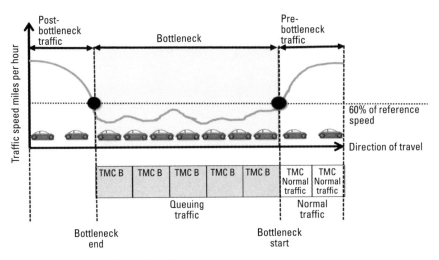

Figure 10.4   Bottleneck speed profile template.

The rightmost black dot on Figure 10.4 indicates the beginning of the bottleneck. This is the point at which the speed of the traffic drops below 60% of the reference speed. The leftmost black dot in Figure 10.4 represents the end, or the point at which traffic climbs through 60% of reference speed.

The gray shaded area in the center of the diagram represents bottleneck conditions where the traffic speed has dropped below 60% and is not yet recovered back to 60% of reference speed. The area to the right of the box represents pre-bottleneck traffic speed, and the area to the left of the box represents post-bottleneck traffic speed. The bottleneck is defined as the width of the gray box.

Along the bottom of the graph in Figure 10.4 a series of boxes is arrayed to represent the TMC segments of the freeway. The two rightmost white boxes are labeled TMC PRB (pre-bottleneck). The small gray boxes under the large gray box are labeled TMCB (bottleneck).

Note that from a data perspective the bottleneck can be defined as a sequence of two pre-bottleneck TMC segments followed by one or more bottleneck segments.

This template provides us with a mechanism for analyzing speed variation to identify and characterize bottlenecks. It also provides the definitions and labels required to explain the analysis.

Making use of the template, an nPath analysis was conducted on the data set to identify bottlenecks, and to characterize bottlenecks by the length of the queue. Figures 10.5–10.7 show several different ways to visualize the results of the analysis.

In Figure 10.5, the size of the circles is proportionate to the duration of the bottleneck measured in minutes below reference speed.

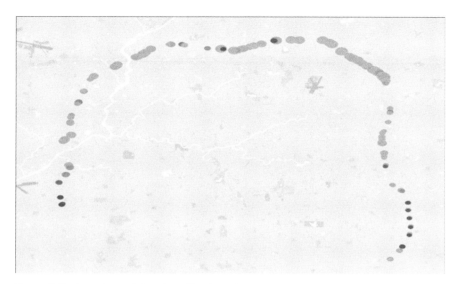

Figure 10.5   Location and duration of bottlenecks.

The data set used to generate the graphic in Figure 10.5 consisted of traffic in both directions over weekday peak periods. Another way to visualize the location of bottlenecks is to identify the nearest major intersection to the bottleneck. Figure 10.6 shows bottlenecks located to major intersections. The data set used to generate this visualization was for all months of 2013 and 2014. The size of the boxes is proportionate to the duration of the bottleneck.

In our last example of a bottleneck visualization, shown in Figure 10.7, the bottlenecks are plotted on a map base of the freeway, with bottlenecks represented by a dot at the location of the bottleneck. The size of the dots corresponds to the duration of the bottleneck in minutes.

The dataset used to generate Figure 10.7 is for the evening peak period for a specific day in a single direction along the freeway. The horizontal axis represents the distance along the freeway, denoted by mile markers. The vertical axis represents the time of day. The size of the dots indicates the duration of the incident in minutes. Therefore a comet trail of bottlenecks stretching back from a higher mile marker to a lower one, as time of day increases, can be interpreted as a chain or sequence of bottlenecks that are probably related. Earlier bottlenecks are likely to be causing later ones. This data was selected to enable the comet trail patterns to be made visible in Figure 10.7. The sequences of bottlenecks illustrate the analysis constraint discussed earlier in this section. Since the template assumes a single dip in speed below the reference speed, the results of the nPath analytics provide a sequence of bottlenecks. It is likely that the comet trails are not a sequence of bottlenecks but, in fact, a single bottle-

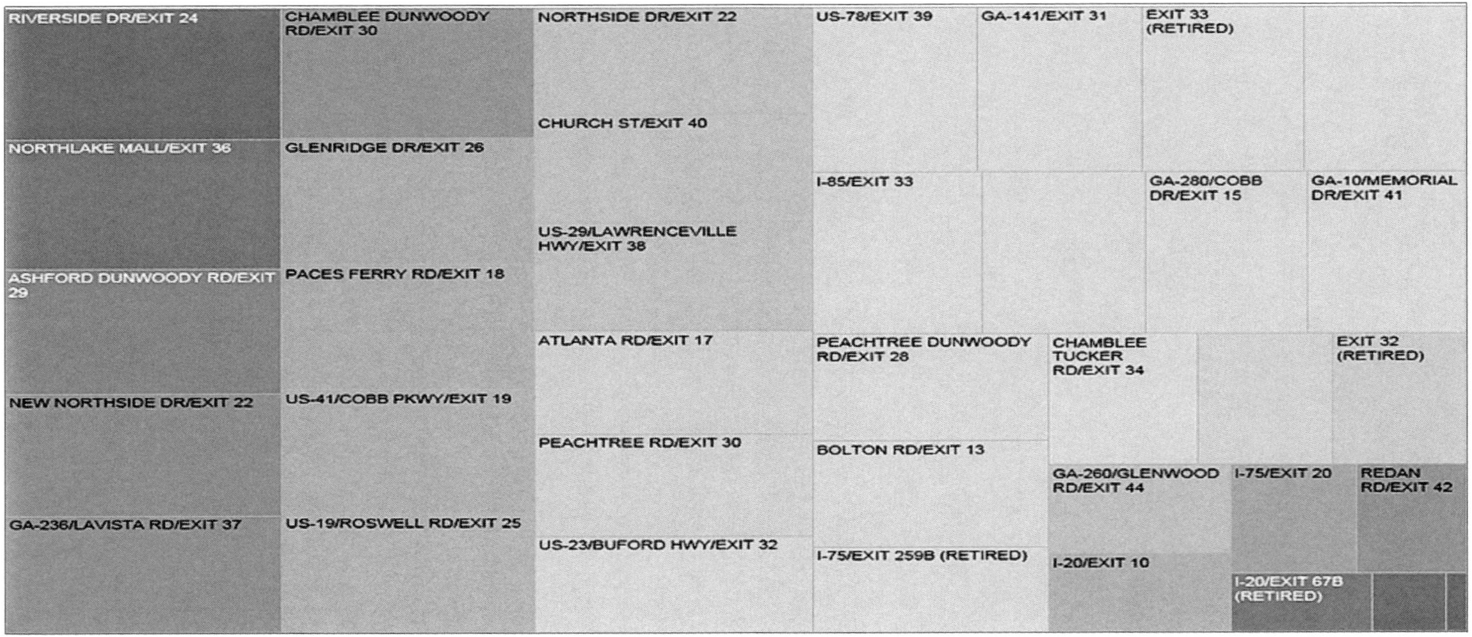

Figure 10.6 Bottlenecks at major intersections with durations.

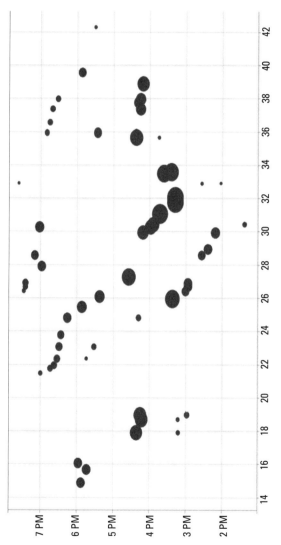

Figure 10.7 Bottlenecks by location and duration for a single day.

neck that is exhibiting multiple speed drops and increases around the reference speed represent a single bottleneck.

Unfortunately, the speed variability, bottleneck, and characterization work discussed above did not yield a statistically significant difference between the conditions before the VSL sign implementation and afterward. This led to discussion and consideration of alternative approaches to the analytics.

Speed variability and bottleneck characterization techniques were applied as part of the original analysis. Unfortunately, this did not yield a statistically significant result that showed an improvement from the before situation to the after situation. Speed variability on its own did not provide a useful measure of the effects of the VSL signs.

At the midway point of the work, some six months into the schedule, a briefing and discussion with the client provided a breakthrough insight into the evaluation problem. In a discussion regarding the subjective effects of the VSL signs, a senior member of the state DOT staff stated:

> When you drive the corridor during incidents conditions, you get the impression that the traffic is somehow more tranquil than before we installed the variable speed limit signs.

This insight led to a review of the parameters that were adopted for the evaluation. It became clear that simple speed variability, standard deviation, or averages may not provide insight addressing the subjective statement. The evaluation team then considered the word tranquil and realized that the opposite of tranquil would be turbulent. This thought led to the determination that an evaluation parameter that measured traffic turbulence may be a more appropriate measurement and might yield a clear result.

Subsequent discussions with the data science team identified a new candidate evaluation parameter—traffic turbulence. Traffic turbulence was defined as the change in speed between adjacent segments times that occurrence of that event. Further analysis also identified that the most significant location to measure traffic turbulence is at the end of the queue, where it would be expected that the VSL effects would be most pronounced due to the warning given to drivers approaching the end of the queue.

This represented a discovery moment. While the previous analytics work did not yield a statistically significant result, it did form the basis for understanding the characteristics of the data sufficient to form the foundation for a traffic turbulence analysis. Given a new understanding of the complexity of the analysis, the team decided to focus on traffic turbulence analysis based on traffic speeds, while putting the other factors to one side.

Revised Approach—Traffic Turbulence Analysis

The team identified that a traffic turbulence analysis would be a more appropriate approach to the evaluation of VSLs. The analysis was also focused on the end of the bottleneck within the zone of influence but beyond the bottleneck. To facilitate the analysis a revised bottleneck speed profile template was created as illustrated in Figure 10.8.

Like Figure 10.4, the vertical axis of the graph in Figure 10.8 represents traffic speed in miles per hour, averaged over one-minute time increments. The horizontal axis represents distance, with the arrow indicating the direction of travel for the traffic. The rightmost black dot on Figure 10.8 indicates the beginning of the bottleneck. This is the point at which the speed of the traffic drops below 60% of the reference speed. The leftmost black dot represents the end, or the point at which traffic climbs through 60% of reference speed. The gray shaded area in the center of the diagram represents bottleneck conditions where the traffic speed has dropped below 60% and has not yet recovered back to 60% of reference speed. The area to the right of the box represents pre-bottleneck traffic speed, and the area to the left of the box represents post bottleneck traffic speed. The bottleneck is defined as the width of the gray box.

Along the bottom of the graph in Figure 10.8, a series of boxes is arrayed to represent the TMC segments of the freeway. The two rightmost white boxes are labeled TMC PRB (pre-bottleneck). The small gray boxes under the large gray box are labeled TMCB (bottleneck).

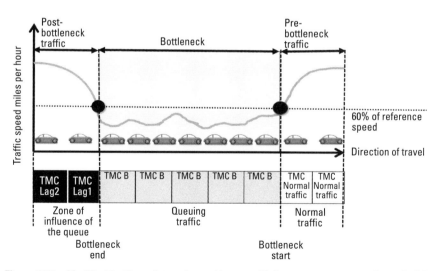

Figure 10.8 Modified bottleneck template with zone of influence segments at the end of the queue.

In Figure 10.8, two additional boxes have been added—the leftmost black boxes that are labeled "TMC LAG1" and "TMC LAG2." These represent post-bottleneck traffic speed conditions.

This post-bottleneck traffic zone can also be defined as the zone of influence of the bottleneck, and the zone speed has been defined as the average of speeds observed in TMC LAG1 and TMC LAG2. The addition of these two segments makes it possible to conduct analysis on the zone of influence with respect to traffic turbulence.

To provide an intuitive measure of the impact of turbulent speed change between adjacent TMCs, we adopted a calculation of acceleration/deceleration and cumulatively calculated the total absolute value of these acceleration/deceleration calculations. The methodology for measuring the acceleration between two adjacent TMC's (x and y) is as follows:

$$Traffic\ turbulence = \left\{ \left[\begin{array}{l} sum\ across\ all\ pairwise\ TMC\ segments\ of \\ \left[\begin{array}{l} \left(\begin{array}{l} TMC\_x\_speed\ * \\ TMC\_x\_speed \end{array} \right) \\ - \left(\begin{array}{l} TMC\_y\_speed\ * \\ TMC\_y\_speed \end{array} \right) \end{array} \right] \end{array} \right] \Big/ \begin{array}{l} Total\ number\ of \\ pairwise\ TMC\ segment \\ comparisons \end{array} \right\}$$

where

TMC x speed is the speed observed on segment X for 1 minute;

TMC y speed is the speed observed on segment Y for one minute;

TMC x pmm is the pseudomile marker for segment X;

TMC y pmm is the pseudomile marker for segment Y.

The measurement units for the above calculation are (miles/hour) per minute

Developing this analytic for peak period traffic on weekdays, using the two post-bottleneck TMC segments, TMC LAG1 and TMC LAG2, yielded the results summarized in Figure 10.9.

Figure 10.9 shows the results of the turbulence analysis for the fourth quarter of 2014 compared to the fourth quarter 2013. As explained earlier the VSL signs were turned at the beginning of the fourth quarter of 2014, defining the after period. The results show a clear reduction in turbulence and speed at the end of the bottleneck. The result was not checked for statistical significance but was of sufficient significance to satisfy the client's needs. Note that there are approximately 71.5 million data points in the turbulence calculation.

Year of Bottleneck Suspected	Lag2 Speed	Lag2 Posted	Lag1 Speed	Lag1 Posted	End Speed	End Posted	Avg. Zone Turbulence	Count of Bottlenecks	Total Zone Turbulence	Avg. Queue length
2013	56.6		49.0		30.1		20.8	81,218	1,686,757	3.9
2014	55.9	47	48.9	43	29.9	39	18.9	80,627	1,521,833	4.3

Figure 10.9 Traffic turbulence calculations for the fourth quarter of 2014 compared to the fourth quarter of 2013.

Toll Return Index Concept

Toll agencies face a specific problem with respect to justifying the level of tolls that users must pay. In many cases, the focus is placed on the amount of the toll to the detriment of a simple explanation of the benefits delivered in return for the toll. The challenge is that the level of toll paid is a simple number, while the benefits delivered may be distributed across multiple departments and multiple budgets within the agency. The toll return index concept attempts to address this challenge by making use of analytics to generate a simple number that represents the benefits delivered in return for the toll. The toll return index is a single number that represents the amount of toll paid divided by the total benefits delivered. The total benefits delivered are comprised of safety, efficiency, and user experience benefits. The concept was developed in close association with a leading toll agency and the relevant professional society [3].

Analytics Used

In this example, the toll return index is composed of four primary components:

- *Toll paid:* There are a number of different ways in which the toll paid could be characterized. In this example, the total tolls paid by a user during an average commute day will be used. This means that the benefits portion of the toll return index will also be calculated in the same basis to determine the index.

- *Safety benefits:* The value of safety benefits is related to the number of crashes avoided through the increased safety on the toll road. The increased safety can be because of better traffic management, better road geometry, or better overall operations. It is determined by comparing crash statistics for the subject freeway to comparable local, regional, and national roads. Crash statistics are also normalized by considering the volume of traffic. Therefore, the crash statistics will be quoted in terms of crashes per vehicle mile traveled. Crash statistics will be obtained from local crash databases. Traffic volumes may be obtained from traffic count data available from the local transportation agency, or from the toll agency itself. In some cases, traffic volumes may be estimated using mathematical simulation models. Crash statistics are typically classified as follows:
  - Fatal;
  - Injury;
  - Damage-only.

According to U.S. DOT statistics, the lifetime comprehensive cost to society for each fatality is $9.1 million, of which 85% is attributable to lost quality of life [4]. Each critically injured survivor has an assumed cost to society of $5.6 million USD on average. Lost quality of life accounts for 82% of this figure [4]. The statistics for injury crashes could be further classified according to injury severity, with different comprehensive cost figures for each subcategory. The previous data for critically injured survivors is the highest level of severity. These can be applied to the crash savings to determine an estimate of the safety benefits.

- *Efficiency benefits:* Efficiency benefits are derived primarily from time savings attributable to the toll road providing a faster route than parallel alternatives. Efficiency benefits are evaluated by comparing time required to travel on the specific road segments, to the time required to travel on alternative routes. The data can be obtained from journey time surveys of both the target road segments and the alternative routes or from a regional transportation model (mathematical simulation model). Another factor that can be considered in the determination of efficiency benefits is the variability or reliability of the trip time on the toll road compared to parallel alternative routes. In the case of speed variability, the average speed variability for each trip using the toll road would be compared with an equivalent trip using alternative routes. The difference in variability between toll road trips and parallel road trips will be quoted in minutes. If the reduction in variability or an increase in reliability will unlock extra minutes for the driver, a value of time calculations can then be made. Once the time savings and reliability improvements have been determined, an estimate of the value of time can be applied to determine a dollar figure for the value of efficiency benefits, either as a function of absolute travel time savings, or also incorporating travel time reliability as a factor.

- *User experience enhancement benefits:* Determining the value of user experience enhancement benefits is the most challenging of the three components. User experience is subjective and difficult to measure. The proposed approach involves the use of social media analysis, which, in the case of this example, is the analysis of a Twitter feed. The Twitter feed is processed to include only those tweets that relate to the target road segments and the alternative routes. Context-specific text analytics are conducted to identify positive and negative comments for users of both groups of road segments. It is then necessary to apply a dollar value to different levels of positive and negative comments. Research on the subject was unable to identify published data regarding the value of public perception; therefore, the toll agency would have to develop specific values for the agency. As an example, if user perception is measured on a scale from 0 to 100% for both the road in question and parallel alternatives, then a positive difference in perception of 1% could be valued at $10,000. This is an arbitrary number intended to provide an example of the calculation. In practice, the value could be calculated as a percentage of the total expenditure of the toll agency in marketing and outreach. For example, if the toll agency spends $5 million per year on marketing and outreach, then a proportion of this figure could be used to represent the value of a 1% improvement in user perception.

Another approach to establishing the value of a 1% change in user perception would be to conduct direct surveys on the user population. The survey would ask users to place a dollar value on the improved experience delivered by the toll road. Smart phone apps could be used to make this an efficient and continuous process.

10.6   Smart City Accessibility Index

Many of the objectives related to smart city initiatives center on improving accessibility to jobs, education, health, and retail opportunities. This involves the measurement of the ease or difficulty of travel between residential zones within the smart city region and zones that contain such opportunities. In close cooperation with movement analytics data providers, smart city analysts and smart city practitioners, the following concept was developed to address the needs of accessibility analysis within a smart city. Typically, transportation accessibility has been defined in urban areas by making use of synthetic data from transportation land-use models. These take relatively small samples of real transportation conditions and apply modeling techniques to develop a big picture for prevailing and future conditions. With the advent of movement analytics from smart phone apps, it is possible to revisit the approach and define a new approach based on observed data. Movement analytics involves the capture of GPS data from smart phones in an aggregated and anonymized manner that enables patterns of travel to and from zones to be determined at a relatively high sample rate. In addition to providing an assessment of overall demand between zones in the smart city region, movement analytics can also provide a strong indication of the modes and routes that are chosen to make the trip. Through the definition of an accessibility index, which is comprised of travel time, travel time reliability, and cost of travel between major zones in the smart city region, it is possible to evaluate accessibility. Note that the movement analytics data also enables the identification of residential zones and those that contain jobs, education, healthcare, and retail opportunities. Movement analytics data can also be blended with U.S. census data to provide additional insight into the characteristics of each zone. While this approach still contains an element of estimation, it is based on observed data that can consider the dynamics of accessibility over time.

Analytics Used

The primary analytic used is known as the accessibility index and it is comprised of the following components:

- *Estimated travel time between residential and opportunity zones:* Zones within the smart city region are defined, based on U.S. census tract boundaries as census data will be required to define demographics for each zone. Movement analytics from smart phone applications will be used to characterize travel between zones and to identify overnight residential locations for smart phone users. Opportunity zones will be identified based on U.S. census data. Travel time analytics will be determined to illustrate average travel times between residential zones and opportunity zones. This analytic will be a factor in the overall accessibility index.

- *Estimated travel time reliability between residential and opportunity zones:* By considering travel times between zones over a suitable period it is possible to develop an estimate of travel time reliability between residential and opportunity zones. This analytic will also be a factor within the overall accessibility index.

- *Cost of travel between residential and opportunity zones as a proportion of household income:* U.S. census data would be used to estimate household income for each residential zone. Cost of travel between residential and opportunity zones will be determined based on the travel time and travel time reliability between zones and an assumed hourly cost of travel. This latter figure is typically related to the average earnings per hour for the city, but it could also consider the average earnings per hour for each zone. The three analytics are combined into a single accessibility index, which is defined as: accessibility index from zone 1 to zone 2 = travel time between zones, travel time variability between zones, and the cost of travel as a proportion of household income between zones. Note that it would also be possible to apply the type of generalized cost modeling used in transportation simulation models as an alternative to this approach.

10.7   Arterial Performance Management

Arterial performance management has become a very important subject area within the transportation industry. Federal government legislation has placed an emphasis on results-driven investments for transportation service delivery. This makes arterial performance management a valuable analytics concept for a smart city. The nature of arterial performance management analytics is described in the following section.

Overview

The basis of the concept is to take the technique applied to freeway management as described in Section 10.6 and apply it to arterial performance management. Freeway and arterial performance environments vary considerably with respect to average traffic speed and the expected variation in speed. Arterial roads are subject to additional factors of complexity, such as conflict between mainline traffic sites, rate traffic, and from pedestrians. Also, the spacing between intersections in arterials is expected to be considerably less than that of freeways, and while a freeway can be treated as a linear corridor, arterials require treatment as a network of nodes and links. To address arterial performance management, the same traffic turbulence analytic developed for freeways would be applied to characterize the severity of acceleration and deceleration being experienced by traffic along the length of the arterials. This could also be supplemented by social media analysis to provide an additional perspective on driver perception with respect to arterial performance management.

Analytics

As discussed earlier, the traffic turbulence analytic would be the primary tool to be used to conduct the evaluation on arterial performance management. This would be combined with traffic flows and an analysis of traffic speed variability along the length of each segment to determine the technical performance of each arterial. A user sentiment index would also be used, based on context-specific keyword analysis of Twitter feeds to determine the percentage satisfaction of drivers along the arterial corridor.

10.8   Decision Support for Bus Acquisition

This concept was developed in discussion with a leading transit agency [5]; the challenge of selecting the appropriate timing for bus acquisition was identified. The purchase of buses to support regional and local bus services represents a significant investment. Each bus can cost $300,000–600,000 depending on the exact specification and the choice of bus. This represents a significant capital investment, placing significant value on any tools or insight that can make such investments as cost-effective as possible. It was identified that one of the major factors in the decision to buy new buses related to the traffic conditions experienced by buses on the roads along which the routes are operated. The transit agency in question operates approximately 1,000 buses in the suburban area of a major city. Consequently, many of the routes operate on arterials and other roads with traffic signals. The overall concept for the use of the analytics support approach to bus acquisition is to characterize prevailing traffic conditions as experienced by the bus fleet, by applying analytical techniques to traffic flows. In this case, rather than focusing on traffic turbulence, the average travel time and travel time variability experienced by individual buses on bus routes is the subject of the analysis. The approach is to develop an index that combines travel time and travel time variability, along with data regarding schedule compliance, to identify the optimum point at which bus acquisition should be considered. It should also be noted that the use of adaptive coordinated traffic signal management techniques, bus priority at intersections, contra flow bus lanes, and special bus right-of-way could also be effective strategies to improve the conditions encountered by buses and bus service quality levels along corridors, in addition to the acquisition of new buses.

Analytics

It is envisaged that the following analytics will be used in the development of a bus acquisition support tool:

- Bus travel speeds per route;
- Individual bus schedule compliance;
- Average traffic speeds per route;
- Bus travel speed variability per route;
- Acceptable bus schedule compliance variation;
- Cost of bus schedule compliance variation.

10.9 Thoughts on the Use of Analytics

A few additional thoughts are offered in each of the five concepts, in terms of the use of the analytics and the insight that can be achieved.

Freeway Management

During the analytics work several challenges were encountered. Some of the challenges related to the building of a bridge between data science and transportation. The effort required in building such a bridge and bringing data scientists up to speed, so to speak, on the characteristics of traffic data was underestimated. Similarly, the effort required to communicate data science to transportation professionals was also underestimated. Other challenges related to the discovery nature of the undertaking. The whole process of applying a discovery tool to a data set, by its very nature means that things that were not perceived at the beginning of the process can become important. It is also the case that one discovery leads to another. This was certainly the case in this exercise, and considerable effort was expended on the initial analysis of a significant extension in the originally envisioned schedule for the work. The original work schedule spanned a period of approximately two months, when, in fact, the actual work spanned a period of more than 12 months.

Another challenge lay in the definition of TMC segments. These vary in length from 500m to more than 2,000m. This limits the resolution of the evaluation, as effects can only be analyzed over the length of the segment and not within the segment. Since the work was completed, INRIX [1] has introduced a new data set with shorter and more consistent segment lengths. This supports a high-resolution of analysis and enables the possibility that this data source could be used on arterials where the distances between intersections and the variability in traffic speeds is greater. The technique developed during the study will also be extremely valuable when used in conjunction with the connected vehicle data. Connected vehicle data offers the possibility of second-by-second speed profiles emanating from connected vehicles. This high-resolution data set could be utilized with the techniques developed here to provide greater insight into the driver behavior at the beginning, during, and at the end of a bottleneck.

A significant discovery element of the project was a realization that the analysis could form the basis for a new scientific approach to traffic engineering. As more data becomes available and as the accuracy and resolution of the data grows, the principles revealed in this project could be applied to the adoption of a scientific approach to traffic engineering.

This would be based on a detailed understanding of the variations in traffic conditions and on the detailed effects of traffic management tools and devices. With the advent of connected vehicle technology, which would enable instantaneous vehicle speed, vehicle location, and vehicle ID to be gathered on a large scale, it is likely that the ground will be prepared for a scientific approach to traffic engineering. It is our belief that connected vehicle data will need to be incorporated into an integrated transportation data set that includes crashes, incidents, road geometry, roadmaps, road signs, weather conditions and other data on which discovery techniques can be applied to support a scientific approach.

Scientific traffic engineering would ultimately lead to the design of traffic-management strategies driven by data and effects. One could also ultimately foresee a day when design documents could be rewritten to take account of this new scientific approach. In parallel with the adoption of scientific traffic engineering, it would also be possible to take a scientific approach to the planning for future investments in transportation facilities including intelligent transportation systems. Access to an integrated data set covering multiple aspects of transportation service delivery and prevailing conditions would also support before studies to be conducted at any point in the span of the data set. Identifying and understanding the exact effects of investments made in the past would lead to finely targeted investments in the future. This could well be the genesis of a new scientific approach to transportation planning and engineering.

A thought related to the scientific approach lies in the need to define, from a data perspective, terms that are typically used by transportation people on a subjective basis. These include terms like recurring and nonrecurring congestion and bottleneck, as illustrated in this chapter.

A final thought related to the freeway speed variability work concerns the availability of performance data beyond the boundaries of the agency. The availability of such data sources could put pressure on local agencies regarding operational efficiency. As an example, during the freeway work the data science team noted the occurrence of an incident on September 20, 2014 between 3 and 4 p.m. In discussion with the client team, it turned out that this was the time and date of a major snowstorm. The significance in this interaction lies in the fact that an independent data science team could identify performance issues with no recourse to public agency data.

Accessibility Index

An interesting aspect of the accessibility analytic lies in the ability to use observed data where typically synthetic data from a transportation model would be used. There is a possibility that the observed data could be combined with this synthetic data to provide a more complete picture of the smart city. However, it would be necessary to consider the accuracy of the resultant data set, since data from the transportation model is likely to be based on a much smaller sample size than the observed data.

Toll Return Index

Once the toll return index framework has been established, it is possible to use in-depth analytics to understand how changes in safety, efficiency, and user experience will affect the toll return index. While the toll return index has been calculated for an entire toll network in this example, it would also be possible to explore the dynamics of the toll return index for different segments of the total network to understand the variations in safety, efficiency and user experience alongside the variation in toll revenue. The toll return index could also be a useful tool in the analysis of variable tolling techniques. In such approaches, the toll is varied dynamically to achieve a specific level of service objective. In many cases, the level of service objective is an average speed of 45 mph over the variable tolling section of the network. The toll return index could be used as an objective function in the calculation of the toll to be charged any given time. Current dynamic tolling implementations rely on a lookup table that compares traffic volumes and speeds with different toll levels. This typically takes no account of factors such as trip purpose when determining the elasticity associated with the variable tolling deployment (with elasticity being defined as the number of vehicles that are redirected from the toll road to alternative parallel rights for each incremental increase in the toll level).

Arterial Performance Management

The incorporation of user perception using social media analysis is likely to yield some interesting results. While traffic and transportation engineers optimize based on the total delay along the corridor, discussions with drivers indicate that they might hold the number of stops along the corridor as more important than the total delay. This seems to be because drivers won't mind a slightly longer journey time along the corridor provided it is a smoother journey with less stops. This would suggest that a lower average speed would be acceptable to drivers if the number of stops were minimized.

Decision Support for Bus Acquisition

While this analytic has been described in terms of a tool to enable better decisions regarding bus acquisition timing, it also forms part of a larger picture involving the use of scientific investment planning, supported by data and analytics. The principles involved in decision support for bus acquisition could equally be applied to other infrastructure elements that support transportation service delivery within a smart city.

References

[1]   www.inrix.com .The data set for the freeway speed variability study was provided by IN-RIX Inc. and preprocessed by Appian Strategies Inc. before delivery to the data science team.

[2]   Workshop session with Albert Yee, Emergent Technologies, and founder of the Caltrans Freeway Management Academy.

[3]   Discussion with Randy Cole, executive director for Ohio Turnpike, and a toll analytics group under the auspices of the International Bridge Tunnel and Turnpike Association.

[4]   *The Economic and Societal Impact of Motor Vehicle Crashes*, 2010 (Revised), U.S. Department of Transportation, National Highway Traffic Safety Administration, May 2015.

[5]   Discussion with Michael Bolton, deputy executive director, PACE Bus Chicago.

11

Benefit and Cost Estimation For Smart City Transportation Services

11.1 Informational Objectives

This chapter aims to do the following:

- Explain a conceptual framework for estimating benefits and costs associated with smart city transportation services.
- Illustrate assumed configurations for subsystems that support the delivery of smart city transportation services.
- Describe examples of cost estimation for smart city transportation services.
- Provide examples of benefits estimation for smart city transportation services.
- Explain the basis for planning and high-level screening evaluation of smart city transportation services.

11.2 Chapter Word Cloud

Figure 11.1 presents a word cloud that provides an overview of the contents of the chapter.

Figure 11.1   Word cloud for Chapter 11.

11.3   Introduction

This chapter explores the costs and benefits that are associated with smart city implementations. The subject is approached in a balanced manner to achieve benefit and cost guidelines that are of practical value, while managing an important issue. This is a challenging subject as detailed costs and benefits are typically identified as a result of a detailed design process. This detailed design process customizes the implementation to a specific city and a specific set of circumstances. In the design process, technology choices are made that shape the intended deployment in a specific direction that is best suited to the city in question. To determine life-cycle costs that include both capital and operating costs, it is also necessary to understand operating and maintenance costs for the services. This chapter aims to provide general guidance that should be applicable to any smart city.

This book advocates a service evolution approach to determining starting points and identifying the roadmap for smart city deployment. Continuing this theme, the 16 smart city services identified in Chapter 5 can also form the basis for a benefit-cost sketch planning framework for smart city transportation. Such a model can provide the high-level guidance required to shape and influence investment patterns for a smart city. It is also recognized that the analytical approach defined in Chapter 5 can be applied to refine benefit and cost estimates for smart city services. Therefore, the cost and benefit model described in this chapter can also be viewed as a starting point or platform that can be refined over time as benefit and cost data from smart cities is collected, and analytics that characterize these are developed. While the benefit-cost framework defined in this chapter has immediate and independent value, it is likely that the major elements and its value will ultimately lie in setting the foundation for future enhancement and development. The intention is to describe the framework and the overall approach to estimating benefits and costs for smart city transportation services, then continue to improve the model incrementally using such analytic techniques. The goal is to develop an advanced decision-support system for smart city service evolution. This will take account of smart city policy objectives and provide guidance on what services should be deployed and the order in which the deployment would take place to optimize several factors. These include benefit-cost ratios, incorporation of legacy systems, and in nature of prior investments. The definition of an optimum sequence of services sets the scene for detailed prioritization and staging and the delivery of services through the definition and implementation of projects. As noted earlier, projects are essential since they represent manageable units of deployment that can be subject to schedule and cost control and represent a standalone step toward future big picture. However, it must be recognized that value and benefits are in fact delivered by the services that are enabled by the projects. Making use of services to communicate the plan investment pattern also enables more effective outreach to a range of smart city partners and to the public, or end users of the proposed services.

Another important goal in defining a benefit-cost framework is to provide guidance to smart cities on the data required for effective before-and-after evaluations. In prior investments in advanced transportation technology, before-and-after data collection has been inconsistent, presenting a challenge to direct comparisons of before-and-after situations and comparisons between cities [1].

11.4   Overview of the Approach

Figure 11.2 illustrates an overall approach to the estimation of benefits and costs for smart city transportation services. Each step is described in the following sections.

Define Smart City Transportation Services

The starting point in the methodology is to create a definition of the smart city transportation services to be deployed. While these are likely to vary from city to city the 16 services identified in Chapter 5 are used to explain and illustrate the general approach. The 16 services are as summarized in Tables 11.1–11.3. Note that these are copies of tables from Chapter 5, showing the objectives mapped to the services. There are two reasons for this. First, objectives provide some insight into the types of values and benefits that may be delivered by the service, and second, it is desirable to reinforce the connection between services and objectives. The primary goal in delivering the services is to achieve the previously defined objectives. The objectives are also grouped according to safety, efficiency, and user experience objectives.

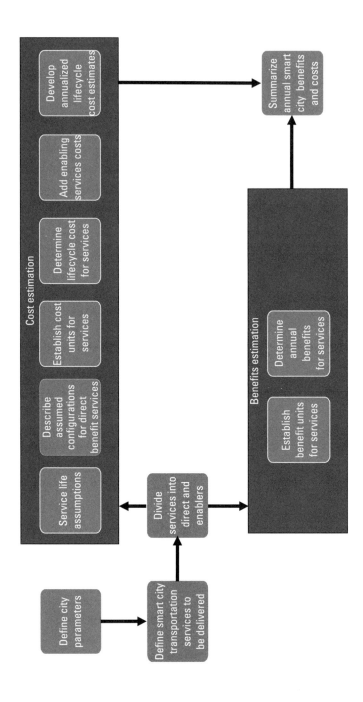

Figure 11.2   Cost and benefit estimation methodology.

Table 11.1
Smart City Transportation Services Mapped to Safety Objectives

Objectives	Services															
	Asset and maintenance management	Connected vehicle	Connected, involved citizens	Integrated electronic payment	Intelligent sensor-based infrastructure	Low cost efficient, secure and resilient ICT	Smart grid, roadway electrification and electric vehicle	Smart land use	Strategic business models and partnering	Transportation governance	Transportation management	Traveler information	Urban analytics	Urban automation	Urban delivery and logistics	User focused mobility
Safety																
Maximizing the safety of the overall transportation system	●	●	●		●	●	●		●	●	●	●	●	●		
Balancing safety, efficiency, and user experience to create a holistic approach									●	●	●	●	●	●		
Understanding the effects of safety improvements and investments													●			

Separate Services into Those That Deliver Direct Benefit and Those That Are Enablers

The 16 services can be divided into those that deliver direct value and benefits and those that act as enablers to allow the other services to deliver the value. This does not suggest that the enabling services have no value; rather it indicates that their value is closely related to the support of the other services.

Describe Assumed Configurations for Direct Benefit Services

While detailed designs are not available when considering a generic smart city, it is necessary to describe the high-level configurations that are assumed for cost and benefit estimation purposes. These can be likened to high-level system architectures that describe the overall components and explain the basis for the benefits and cost estimates.

Table 11.2
Smart City Transportation Services Mapped to Efficiency Objectives

Objectives	Asset and maintenance management	Connected vehicle	Connected, involved citizens	Integrated electronic payment	Intelligent sensor-based infrastructure	Low cost efficient, secure and resilient ICT	Smart grid, roadway electrification and electric vehicle	Smart land use	Strategic business models and partnering	Transportation governance	Transportation management	Traveler information	Urban analytics	Urban automation	Urban delivery and logistics	User focused mobility
Efficiency																
Identifying bottlenecks and slowdowns in the transportation system and developing strategies to manage these		●	●	●	●	●					●	●	●	●		
Optimizing the performance of transportation assets and infrastructure	●				●	●			●				●	●		
Optimizing the performance of the overall transportation system									●	●	●		●	●		
Minimizing capital and operating costs and optimizing current and future expenditure	●							●	●	●		●	●	●		
Improving service levels for citizens and visitors	●	●	●	●	●	●	●	●	●	●	●	●	●		●	
Identifying and addressing service deficiencies in the transportation system				●	●	●					●	●	●			
Assessing current and future demand for transportation					●	●		●			●	●	●			
Optimizing access to jobs								●			●	●	●	●		●
Optimizing transportation service for the transportation-disadvantaged		●														
Understanding the relationship between cost of travel and demand for travel													●			

Table 11.2 (continued)

Objectives	Asset and maintenance management	Connected vehicle	Connected, involved citizens	Integrated electronic payment	Intelligent sensor-based infrastructure	Low cost efficient, secure and resilient ICT	Smart grid, roadway electrification and electric vehicle	Smart land use	Strategic business models and partnering	Transportation governance	Transportation management	Traveler information	Urban analytics	Urban automation	Urban delivery and logistics	User focused mobility
Enhanced user experience																
Understanding current user experience			●	●	●	●							●	●	●	●
Developing strategies and techniques for optimizing user experience									●	●	●		●	●		
Delivering the highest value for money for all transportation customers	●			●					●		●		●	●	●	●
Identifying customer perception of current transportation service			●							●	●		●	●	●	●
Assessing the effect of transportation as a service on the overall transportation system									●				●		●	●

Establish Cost Units for Services

As described in Chapter 5, the services will evolve over time and space. In this case, space represents the geographic extent or the proportion of the city population that is exposed to the services at any given time.

Establish Benefit Units for Services

In a similar manner to the cost units described above benefit units can be determined for each service. The benefit can be related to space, time, and quality of service.

Table 11.2 (continued)

Objectives	Asset and maintenance management	Connected vehicle	Connected, involved citizens	Integrated electronic payment	Intelligent sensor-based infrastructure	Low cost efficient, secure and resilient ICT	Smart grid, roadway electrification and electric vehicle	Smart land use	Strategic business models and partnering	Transportation governance	Transportation management	Traveler information	Urban analytics	Urban automation	Urban delivery and logistics	User focused mobility
Enhanced user experience																
Understanding current user experience			●	●	●	●							●	●	●	●
Developing strategies and techniques for optimizing user experience									●	●	●		●	●		
Delivering the highest value for money for all transportation customers	●			●					●		●		●	●	●	●
Identifying customer perception of current transportation service			●							●	●		●	●	●	●
Assessing the effect of transportation as a service on the overall transportation system									●				●		●	●

Determine Life-Cycle Cost for Services

Life-cycle costs represent a combination of the capital cost and operating costs required to deliver the services. Both are likely to have a direct relationship to the spatial extent and quality of the service.

Table 11.3
Smart City Transportation Services Mapped to User Experience Objectives

Objectives	Asset and maintenance management	Connected vehicle	Connected, involved citizens	Integrated electronic payment	Intelligent sensor-based infrastructure	Low cost efficient, secure and resilient ICT	Smart grid, roadway electrification and electric vehicle	Smart land use	Strategic business models and partnering	Transportation governance	Transportation management	Traveler information	Urban analytics	Urban automation	Urban delivery and logistics	User focused mobility
Enhanced user experience																
Understanding current user experience			●	●	●	●							●	●	●	●
Developing strategies and techniques for optimizing user experience									●	●	●		●	●		
Delivering the highest value for money for all transportation customers	●			●					●		●		●	●	●	●
Identifying customer perception of current transportation service			●						●	●			●	●	●	●
Assessing the effect of transportation as a service on the overall transportation system									●				●		●	●
Ensuring that travelers make the best use of the transportation system, taking account of transportation demand and prevailing conditions			●	●	●	●			●	●	●	●	●	●	●	
Understanding the relationship between high-quality traveler information and travel behavior			●	●	●	●			●	●	●	●	●			

Determine Benefits for Services

Service benefits are derived from prior implementation experience, assuming that a similar level of benefit will be derived for the smart city as experienced in previous deployments. As there are only a small number of prior smart city deployments, previous deployments include intelligent transportation system deployments as well as smart city deployments.

Summarize Benefits and Costs

The final step in the process is to summarize the benefits and costs in preparation for using these figures as input to decision support for service delivery sequencing

11.5 Assumptions

Even at a sketch planning level, it is necessary to make some overall assumptions regarding the size and nature of the smart city as the basis for a cost and benefit estimation. A model smart city has been defined and can be described by the parameters listed in Tables 11.4–11.6. The numbers have been selected to be as realistic as possible, but as noted earlier, detailed cost and benefit estimates require a detailed design exercise. Table 11.4 summarizes U.S. national parameters used in the cost estimation.

These figures were used as the basis for cost estimation. U.S. population statistics were obtained from census data [2]. National figures for gas stations [3] were proportioned according to the U.S. population.

Private car operating costs and urban truck operating costs were obtained from the AAA website [4], assuming that an urban delivery truck would have

Table 11.4
U.S. National Parameters Used to Model the Smart City

USA National Parameters	
U.S. population	324,838,897
Gas stations in the United States	123,807
Population per gas station	2,624
Private car operating costs per mile	$0.580
Urban truck operating costs per mile	$0.708
Private car emissions costs per mile	$0.011
Urban truck emissions costs per mile	$0.026
Electric vehicle operating cost per mile	$0.140
Value of time in dollars per hour	$12.5
Average time spent traveling per person per day in minutes	79.5
Total cost of ownership of 1 TB of data per year	$12,000

Table 11.5

Estimation of the Number of Devices in the Smart City

Number of Devices in the Smart City	Miles	Devices per Mile	Total Devices	Notes
Length of smart city freeways	51	4	204	One vehicle detection station and one dynamic message sign in each direction per mile
Length of smart city arterials	126	2	252	One intersection per mile, one traffic controller, and one vehicle detection system per intersection
Length of smart city urban surface streets	989	4	3956	One intersection every half mile, one traffic control and one vehicle detection system for each intersection
Length of smart city bus routes	372	4	1488	No LRT or commuter rail Two devices per mile in each direction along each route
		Total devices	5,900	

Table 11.6

Smart City Characteristics

Smart City Characteristics	
Addressable traveling population in the smart city	1,000,000
Number of buses in the smart city	539
Number of private cars in the smart city	807,990
Number of taxis in the smart city	7,200
Number of rental cars in the smart city	7,121
Number of urban delivery vehicles in the smart city	80,799
Annual vehicle miles traveled per capita	9,442
Proportion of VMT by private car	88%
Proportion of VMT by truck	12%
Deliveries per day in the smart city	5,000
Number of parking spaces in the smart city	323,237
Number of smart factories in the smart city	200
Number of smart educational facilities in the smart city	200
Number of smart healthcare facilities in the smart city	200
Electric vehicle charging points in the smart city	381
Average cost of urban package delivery in the smart city	$10.00
Electric vehicle operating cost reduction in the smart city	76%
Electric vehicle emissions reduction in the smart city	100%
Fatalities per 100,000 population per year	10.69
Injured persons per 100,000 population per year	752
Average data volume per year in TB	365

similar operating costs to the four-wheel-drive SUV at the highest end of the AAA cost spectrum. This may be conservative, as the number of stops incurred by an urban delivery truck may well increase operating cost. Cost of emissions for private cars and urban trucks were obtained from nationally published figures [5]. The value of travel time in dollars per hour [6] and the average time spent traveling per person per day [7] were derived from published data. The total cost of ownership of 1 TB of data per year was derived from a published white paper [8].

Table 11.5 illustrates the estimation of the number of devices in the smart city. Notes are provided to explain the assumptions on which the estimates are based.

The basis on which these numbers were derived is discussed as follows. For freeways, arterials and urban surface streets, per capita estimates were developed from national figures [9] and proportioned for a smart city population of 1 million people. Bus routes were derived from published data from the Chicago Transit Authority website [10], proportioned according to population.

Table 11.6 provides details of smart city characteristics used in cost estimation. These have either been assumed or derived from the national data shown in Table 11.4.

The addressable population is assumed to be 1 million people. Note that this does not include the very young and the very old and is assumed to represent the traveling population. Number of buses was derived from published data from the Chicago Transit Authority regarding the number of buses and population served, proportioned according to population. This is likely to create a conservative estimate as CTA makes use of commuter rail and may be less dependent on buses than a typical city. CTA was used as a data source as full-service details regarding the number of buses, route miles, and population served are available in a single table on the CTA website [10]. It is assumed that the model smart city does not feature light rail transit or commuter rail.

The number of private cars within the smart city was derived from national figures [11] proportioned according to the smart city population.

Annual vehicle miles traveled per capita [12], proportion of VMT by private car, and proportion of VMT by truck were derived from national figures [13], proportioned according to the smart city population.

The number of urban delivery vehicles in the smart city is assumed to be an additional 10% of the private car total.

The number of deliveries per day in the smart city is a conservative estimate.

The number of taxis in the smart city is derived from national statistics [14], proportioned according to population.

Rental cars and parking spaces were also subject to the same treatment using national figures for rental cars [15] and parking spaces [16], proportioned according to population.

The numbers for smart factories, educational facilities, and healthcare facilities are based on a best estimate of one facility for every 5,000 population.

The average cost of urban package delivery was assumed as a best estimate of $10 USD. With respect to electric vehicle charging points, it was assumed that a smart city will eventually feature the same density of electric vehicle charging points as currently exhibited by gas stations in the United States. National figures for gas stations [3] were proportioned according to the smart city population to derive the number of electric vehicle charging. U.S. population statistics were obtained from the U.S. Census website [2].

The electric vehicle operating cost reduction was derived by comparing published cost of electric vehicle operation, (assuming an electricity cost of $0.10 USD per kilowatt), against the cost of operating a private vehicle [17].

Regarding safety statistics, fatalities per 100,000 population per year and injured persons per 100,000 population per year were derived from nationally published figures [18], proportioned according to the population of the smart city. The assumed societal cost of fatalities and injury accidents were derived from the same reference [18]. With respect to calculations involving the value of travel time saved, the average travel time per person per day [7] was proportioned by the population of the smart city.

With respect to big data and analytics it is assumed that total cost of ownership of 1 TB of data is approximately $12,000 USD per year [8] and that a smart city will generate approximately 1 TB of data every day [19].

11.6 Smart City Cost and Benefit Estimation

Following the methodology explained in the Section 11.5, smart city benefits and cost estimates for each of the 16 services are explained in the following sections.

Define Smart City Transportation Services

As noted in Section 11.5, the 16 smart city transportation services that were defined in Chapter 5 are used as a basis for this methodology. Since several of the services do not provide direct benefit, but act as enablers to release benefit from the other services, it is necessary to separate the services into two categories for the purposes of the methodology. This is described in the next section.

Separate Services into Those That Deliver Direct Benefit and Those That Are Enablers

Table 11.7 shows the 16 services grouped according to direct benefit and benefit-enabling services.

Services are categorized as enablers for the following reasons.

Table 11.7
Smart City Transportation Services Categorized as Direct Benefit or Indirect Benefit Enabler

	Direct	In direct Benefit or Enabler
1	Asset and maintenance management	
2	Connected vehicle	
3	Connected, involved citizens and visitors	
4	Integrated electronic payment	
5		Intelligent sensor–based infrastructure
6		Low-cost efficient, secure, and resilient ICT
7	Smart grid, roadway electrification, and electric vehicle	
8	Smart land use	
9		Strategic business models and partnering
10		Transportation governance
11	Transportation management	
12	Traveler information	
13		Urban analytics
14	Urban automation	
15	Urban delivery and logistics	
16	User-focused mobility	

- *Intelligent sensor–based infrastructure:* These represent the sensors that are installed at the roadside and at waypoints along the transportation system. They provide the data that describes prevailing traffic and transportation conditions that acts as inputs to planning and operational decision-making. As such they deliver no direct value, but play a critical role in service delivery and performance management.

- *Low-cost efficient, secure, and resilient ICT:* These represent the communication network that is used to provide connectivity within a smart city. Data is communicated from sensors, crowdsourcing, and other systems and from connected vehicles to a central back office or a smart city transportation management center for conversion to information, insights, understanding, and action. Such technologies deliver significant value but not in their own right. Their value lies in enabling data to be transmitted and supporting the conversion of data to information.

- *Strategic business models and partnering:* These services focus on the definition of and agreement on business models and partnering arrangements that underpin the efficient delivery of smart city transportation services. A business model defines the following:
  - Revenue sources;
  - Intended customers;

- Services to be delivered;
- Financing and funding arrangements.

A business model should also illustrate the balance between who pays for the delivery of the service and who benefits from any return revenue or profit. Typically, the business model will also form the basis for partnering arrangements that define who does what and under what terms and conditions. Business models and partnering arrangements do not deliver direct value but are essential to the optimization of service delivery in a smart city through the incorporation and convergence of both public- and private-sector resources and motivation.

- *Transportation governance:* To support the effective delivery of transportation services in a smart city, it is necessary to define a government structure that will support the effective planning and operation of the services. The concept of operations can be defined that identifies roles and responsibilities for the various transportation partners involved in the smart city. Here again, this service delivers no direct value, but plays an essential role in optimizing the delivery of transportation services in a smart city
- *Urban analytics:* This is a major subject of the book. Although these measures do not deliver direct value, they act as major enablers in the decision-making process for planning and operations of smart city transportation services. Effective analytics will shape the performance of the smart city from a transportation service perspective in terms of service delivery and efficiency of the overall transportation management and government structure.

Each of the above enabling services will be subject to separate treatment with respect to cost and benefit analysis. It is assumed that each of the services deliver no independent benefits and that the benefits lie in the enablement of the other services. The cost of each service will be derived as a proportion of the cost of delivering the direct benefit services or as a proportion of our overall project scale factor.

11.7 Assumed Configurations for Cost Estimation Purposes

For the 11 direct-benefit services a series of assumed configurations was defined to identify elements of the system that would support the delivery of the service. These are described in the following section.

Asset and Maintenance Management

Three elements have been identified for the system that will support the delivery of the asset and maintenance management service, as illustrated in Figure 11.3. The sensor-based data subsystem element represents the equipment and software required to enable data to be collected from each device being managed in an automated fashion. The manual data input subsystem represents the hardware and software required to enable manual entry regarding device condition and performance. This includes remote terminal and input to a central asset management data and analytics system. The asset management data and analytics element represents a centralized hardware and software system that conducts the functions required for efficient asset and maintenance management for the smart city.

Connected Vehicle

The configuration that has been assumed for the connected vehicle system that will support the delivery of connected vehicle services as described previously and illustrated in Figure 11.4.

There are four essential elements of the connected vehicle system: the private car in-vehicle subsystem, the urban delivery vehicle in-vehicle subsystem, the rental car/taxi vehicle subsystem, and the connected vehicle data and analytics subsystem. All three in-vehicle subsystems are assumed to be similar with respect to cost estimation and consist of the hardware and software required to manage data to and from the vehicle and support the connection from the vehicle subsystem to the connected vehicle data and analytics subsystem. The connected vehicle data and analytics subsystem is a hardware and software environment required to support all functions associated with the data management and analytics regarding connected vehicles.

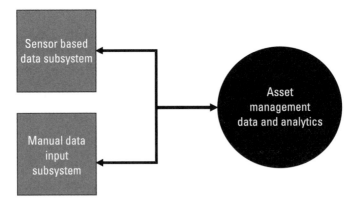

Figure 11.3   Assumed configuration asset and maintenance management service.

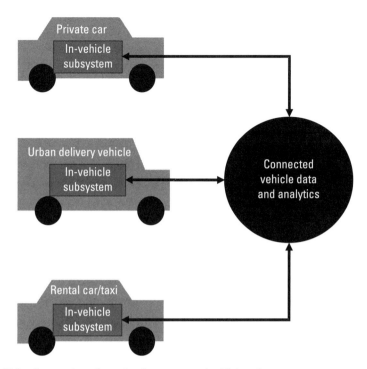

Figure 11.4 Assumed configuration for connected vehicle subsystem.

Connected, Involved Citizens and Visitors

The assumed configuration for the connected, involved citizens and visitors system is illustrated in Figure 11.5. It is comprised of two subsystem elements: the connected citizen and visitor application for a smart phone and the connected citizen and visitor data and analytics subsystem. The connected citizen and visitor application is assumed to be available to both smart city citizens and visitors and consists of a smart phone application that can be installed and operated on a wide range of currently available smart phones. The smart phone application will manage the interface between the citizens and visitors and the connected citizen and visitor data and analytics subsystem. The communication link will support two-way communications with information being delivered to citizens and visitors via the smart phone application and data being collected on a crowdsourcing basis from the smart phone application, on an opt-in basis. The connected citizen and visitor data and analytics subsystem will be comprised of a hardware and software environment capable of supporting all functions required for data management and data analytics associated with connected, involved citizens and visitors.

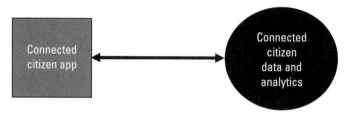

Figure 11.5   Assumed configuration for connected, involved citizen and vistor system.

Integrated Electronic Payment

The integrated electronic payment system that supports the delivery of integrated electronic payment services for the smart city is assumed to consist of five elements: a smart phone payment application, an electronic parking fee collection subsystem, an electronic toll collection subsystem, an electronic ticketing for transit subsystem, and a payment data and analytics subsystem. These are illustrated in Figure 11.6.

The payment applications subsystem is assumed to be a smart phone application that can be used across a wide variety of smart phone devices. The software can support the interface with the user and two-way communications between the payment application and the integrated payment data and analytics subsystem. The electronic parking fee collection subsystem will consist of a hardware and software environment that will support all revenue and management functions associated with electronic parking fee collection for both surface and parking structure spaces within the smart city. These functions will include the collection of data and information delivery with respect to availability and location of spaces. The electronic toll collection subsystem will consist of a hardware and software environment that will support all aspects of electronic toll collection including account management, transponder management, transaction processing, and enforcement. The electronic ticketing for transit subsystem will support similar functions for transit ticketing through purpose-specific hardware and software environment that supports point-of-sale transactions on transit vehicles and back-office processing of data and transactions.

Smart Grid, Roadway Electrification, and Electric Vehicle

Figure 11.7 illustrates the assumed configuration of the system that supports the delivery of smart grid, roadway electrification, and electric vehicle service. There are five elements identified within this system: a private car in-vehicle subsystem, an urban delivery vehicle in-vehicle subsystem, a rental car/taxi in-vehicle subsystem, a roadside electric vehicle charging subsystem, and a smart grid data and analytics subsystem. These are illustrated in Figure 11.7.

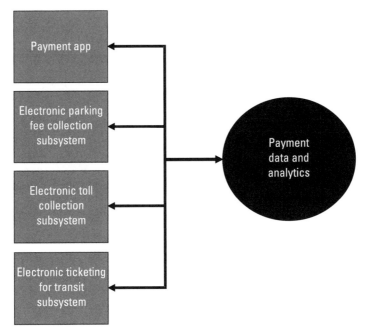

Figure 11.6   Assumed configuration for integrated electronic payment systems.

The in-vehicle subsystems for private cars, urban delivery vehicles, and rental cars/taxis are assumed to be identical and consist of a hardware and software environment that will support management of the electric vehicle charging system and the interaction between the end vehicle subsystem and the roadside electric vehicle charging subsystem. It is assumed that in addition to energy transfer, data transfer is also supported regarding the performance of the vehicle and the battery. This could also be considered as an opportunity to supply the driver with travel information and extract probe vehicle data from the subsystem in the vehicle.

Smart Land Use

Figure 11.8 illustrates the assumed configuration for the smart land-use system.

The smart land-use system is assumed to be comprised of nine separate elements: a private car in-vehicle subsystem, an urban delivery vehicle in-vehicle subsystem, a rental car/taxi in-vehicle subsystem, a smart land-use planning data and analytics subsystem, a movement analytics subsystem, a retail subsystem, a smart factory subsystem, an education subsystem, and a healthcare subsystem. The in-vehicle subsystems are assumed to be identical for all types of vehicles. The smart land-use planning data and analytics subsystem is comprised of a hardware and software environment capable of handling data man-

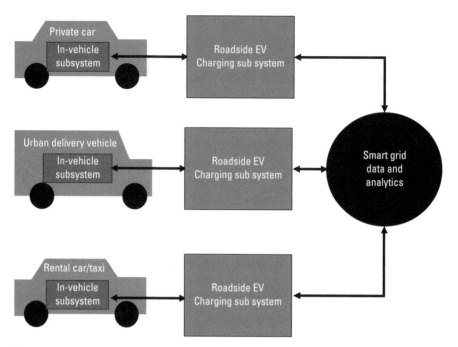

Figure 11.7   Assumed configuration for smart grid, roadway electrification, and electric vehicle system

agement and analytics associated with smart land-use planning. The movement analytics subsystem is a hardware and software environment capable of turning movement analytics data from smart phones into land-use parameters. The retail, smart factory, education, and healthcare subsystems hardware and software environment are designed to support the extraction of relevant land-use data from each for delivery to the smart land-use planning data and analytics subsystem. The intent is that probe vehicle data from the end vehicle subsystems will provide a more detailed and dynamic view of trip generation from various uses within the smart city. The movement analytics data will supplement this data by providing person-specific data to cover other modes of transportation. The retail, smart factory, education, and healthcare subsystems will serve as a bridge between the smart city land-use data and analytics subsystem and internally focused applications designed to improve the retail experience, raise the efficiency of factories, and increase accessibility to education and healthcare services.

Transportation Management

Figure 11.9 illustrates the assumed configuration for the transportation management system that will support the delivery of transportation management services in the smart city.

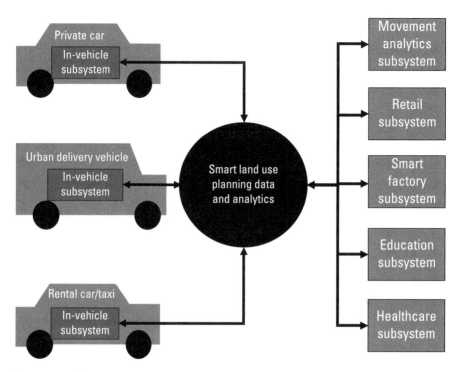

Figure 11.8 Assumed configuration for the smart land-use system.

This configuration is comprised of four elements: a traffic management data and analytics subsystem, a traffic signal management data and analytics subsystem, a transit management data and analytics subsystem, and a smart city transportation management data and analytics subsystem. The traffic management data and analytics subsystem processes data from freeway and incident management sources to supply analytics to the smart city transportation management data and analytics subsystem. The traffic signal management data and analytics subsystem processes data from urban traffic signal control systems and provides analytics to the smart city transportation management data and analytics subsystem; the transit management data and analytics subsystem processes data with respect to prevailing conditions on transit systems and provides analytics to the smart city transportation management data and analytics subsystem. The smart city data and analytics subsystem is a hardware and software environment that supports all functions related to the processing of data and the creation of analytics related to smart city transportation management.

Traveler Information

Figure 11.10 illustrates the assumed configuration for the traveler information subsystem that enables delivery of the associated traveler information service.

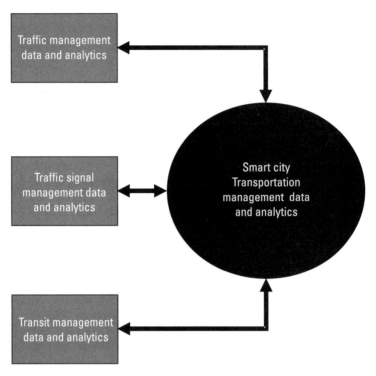

Figure 11.9   Assumed configuration for smart city transportation management system.

Seven elements have been identified: a smart phone apps subsystem, a movement analytics subsystem, a transportation management data and analytics subsystem, a transit management data and analytics subsystem, a freeway management center subsystem, a traffic signal management data and analytics subsystem, and a traveler information data and analytics subsystem. The smart phone application would act as a user interface between travelers and the traveler information and data analytics subsystem. This would provide travelers with current traveler information and enable travelers to provide feedback on prevailing transportation conditions. The movement analytics subsystem would provide GPS-based data from cell phones to enable traveler origin and destination and choice of mode to be determined. The traveler information data and analytics subsystem consist of a hardware and software environment that supports all functions to turn traveler information data into information and to conduct analytics on the data. The transportation management subsystem would act as a data source for the traveler information data and analytics subsystem, along with the transit, freeway, and traffic signal management subsystems.

Urban Automation

Figure 11.11 illustrates the assumed configuration for the urban automation subsystem that enables delivery of the associated urban automation service.

Urban Delivery and Logistics

Figure 11.12 illustrates the assumed configuration for the urban delivery and logistics subsystem that enables delivery of the associated urban delivery and logistics service.

User-Focused Mobility

This is often referred to as MaaS. Figure 11.13 illustrates its assumed configuration.

There are four separate elements defined in the system; some of these are duplicated to indicate that there could be more than one subsystem system in each smart city. The elements identified are a user subsystem, a public MaaS provider subsystem, a private MaaS provider subsystem, and a user-focused mobility data and analytics subsystem. The user-focused subsystem will be a smart phone application that supports the interface between the user and the user-focused mobility data and analytics subsystem. This will support two-way communications with information-available services and cost in one direction and information on user travel needs and the other direction. The MaaS subsystems

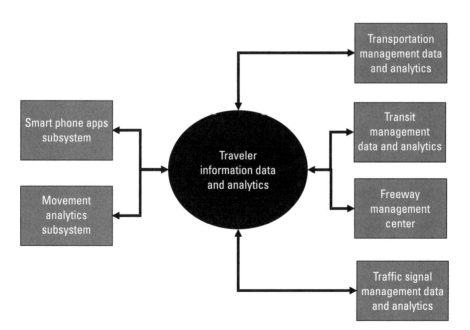

Figure 11.10   Assumed configuration for traveler information data and analytics subsystem.

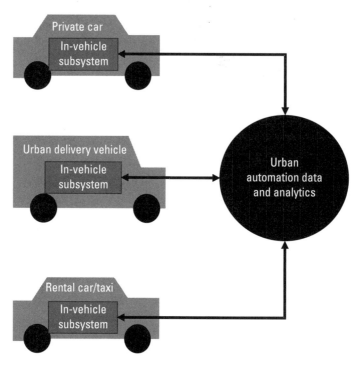

Figure 11.11   Assumed configuration for urban automation system.

will be operated by both public and private enterprises and will support the fleet management and coordination of mobility services and the data collection and information processing required to make a menu of mobility services available to the user. The user-focused mobility data and analytics subsystem will be a hardware and software environment capable of supporting all the functions required to manage MaaS through matching available services to user demand.

11.8   Cost Estimates for Smart City Transportation Services

Making use of the model smart city parameters shown in Tables 11.4–11.6 and the assumed configurations described in Section 11.7, a series of cost estimates was developed for each smart city transportation service. The results of the cost estimation are shown in Tables 11.19–11.24. A commentary has been provided on the cost estimation calculation after each table.

The capital investment required for the asset management central subsystem is assumed to be proportional to the number of devices under management. There are assumed to be a total of 5,900 devices based on the assumptions regarding the number of devices per mile and the number of miles of each road type and bus routes contained in Table 11.8. The number of data devices

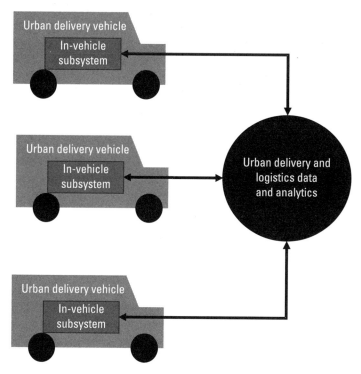

Figure 11.12   Assumed configuration for urban delivery and logistics system.

per mile for each road type and bus route is the best estimate based on the author's experience and represents the average number of devices that are expected to be installed along each type of road and route. A scale factor (capital per unit) of $2,000 USD has been assumed as a best estimate to take account of the relationship between the capital investment required for the central subsystem and the number of devices being managed. Best estimates of $500 USD per unit have also been assumed for the sensor-based data subsystem and the manual data input subsystem scale factors. The former represents additional hardware and software required to automatically extract device performance data from the device, while the latter represents the cost of hardware and software associated with manual data input. A design life of seven years has been assumed for all subsystem elements. Research did not result in a definitive, agreed on lifespan for hardware and software systems, due to the need for judgment regarding the estimated useful lives of hardware and software assets. Given the pace of technological change, it is likely that a hardware and software system will be replaced before the end of its useful life because of advances in technologies and the desirability of new features and functions. The design life of seven years accounts for a technology refresh to be conducted at this point even if the system is still functional. Life-cycle capital costs have been determined by dividing total

Table 11.8

Asset and Maintenance Management Cost Estimate

Asset and Maintenance Management				
Element	Asset Management Central	Sensor Based Data Subsystem	Manual data input subsystem	Totals
Scale factor (capital per unit)	$2,000	$500	$500	
Units	5,900	5,900	5,900	
Total capital investment required	$11,800,000	$2,950,000	$2,950,000	$17,700,000
Annual operating costs	$1,770,000	$442,500	$442,500	$2,655,000
Assumed design life in years	7	7	7	
Annual capital investment required	$1,685,714	$421,429	$421,429	$2,528,571
Annual life cycle cost	$3,455,714	$863,929	$863,929	$5,183,571

Table 11.9

Connected Vehicle Cost Estimate

Connected Vehicle						
Element	Connected Vehicle Central	Private Car In-Vehicle Subsystem	Urban Delivery Truck In-Vehicle Subsystem	Bus In-Vehicle Subsystem	Rental Car And Taxi In-Vehicle Subsystem	Annual Lifecycle Cost Costs
Scale factor (capital per unit)	$50	$500	$500	$500	$500	
Units	903,649	807,990	80,799	539	14,321	903,649
Total capital investment required	$45,182,451	$403,995,000	$40,399,500	$269,500	$7,160,514	$497,006,966
Annual operating costs	$6,777,368	$60,599,250	$6,059,925	$40,425	$1,074,077	$99,401,393
Assumed design life in years	7	7	7	8	7	
Annual capital investment required	$6,454,636	$57,713,571	$5,771,357	$33,688	$1,022,931	$70,996,183
Totals	$13,232,004	$118,312,821	$11,831,282	$74,113	$2,097,008	$145,547,227

capital by the assumed design life of the system. This represents the total capital investment requirement amortized over the design life of the system. Similarly, operating costs have been determined on an annual basis, then both capital life-cycle costs and operating costs are combined to create a total annual life-cycle cost. Annual life-cycle operating costs are assumed to be 15% of total capital costs per year. This assumption has been used for all subsequent cost calculations in this chapter. This aligns with published information [20] that indicates

Table 11.10

Connected, Involved Citizens and Visitors Cost Estimate

Connected, Involved Citizens			
Element	**Connected Citizen Central**	**Connected Citizen App**	**Annual Lifecycle Cost**
Scale factor (capital per unit)	$5	$10	—
Units	1,000,000	1,000,000	—
Total capital investment required	$5,000,000	$10,000,000	$15,000,000
Annual operating costs	$750,000	$1,500,000	$2,250,000
Assumed design life in years	7	1	—
Annual capital investment required	$714,286	$10,000,000	$10,714,286
Total life-cycle costs	$1,464,286	$11,500,000	$12,964,286

Table 11.11

Integrated Electronic Payment Cost Estimate

Integrated Electronic Payment					
Element	**Electronic Parking Fee Collection Subsystem**	**Payment App**	**Electronic Toll Collection Subsystem**	**Electronic Ticketing Collection Subsystem**	**Totals**
Scale factor (capital per unit)	$100	$5	$50	$50,000	
Units	323,237	1,000,000	903,110	539	903,649
Total capital investment required	$32,323,715	$5,000,000	$45,155,501	$26,950,000	$109,429,217
Annual operating costs	$4,848,557	$750,000	$6,773,325	$4,042,500	$16,414,382
Assumed design life in years	7	1	7	7	
Annual capital investment required	$4,617,674	$5,000,000	$6,450,786	$3,850,000	$19,918,460
Annual life-cycle cost	$9,466,231	$5,750,000	$13,224,111	$7,892,500	$36,332,842

that the average annual operating and maintenance cost for a transportation management center is on the order of 15 % of capital cost.

For the connected vehicle central subsystem, a best estimate of $10 USD per unit has been estimated as a scale factor for the cost of the hardware and software. For the private car in-vehicle subsystem, it is assumed that this will add $500 USD to the cost of in-vehicle electronics. This same best estimate is used for urban delivery truck, rental car, and taxi in-vehicle subsystems. A de-

Table 11.12
Smart Grid, Roadway Electrification, and Electric Vehicle Cost Estimate

Smart Grid, Roadway Electrification And Electric Vehicle				
	Smart Grid Management Central	Roadside Electric Vehicle Charging Subsystem	In-vehicle Subsystem	Total
Scale factor (capital per unit)	$10,000	$100,000	$500	
Units	381	381	903,649	
Total capital investment required	$3,811,335	$38,113,354	$451,824,514	$493,749,204
Annual operating costs	$571,700	$5,717,003	$67,773,677	$74,062,381
Assumed design life in years	7	7	7	
Annual capital investment required	$544,476	$5,444,765	$64,546,359	$70,535,601
Annual life-cycle cost	$1,116,177	$11,161,768	$132,320,036	$144,597,981

Table 11.13
Smart Land-Use Cost Estimate

Smart Land Use						
Element	Smart Land Use Central	Movement Analytics Subsystem	Smart Factory Subsystem	Smart education subsystem	Smart healthcare subsystem	Totals
Scale factor (capital per unit)	$10	$10	$50,000	$50,000	$50,000	
Units	903,649	1,000,000	200	200	200	
Total capital investment required	$9,036,490	$10,000,000	$10,000,000	$10,000,000	$10,000,000	$49,036,490
Annual operating costs	$1,355,474	$1,500,000	$1,500,000	$1,500,000	$1,500,000	$7,355,474
Assumed design life in years	7	7	7	7	7	
Annual capital investment required	$1,290,927	$1,428,571	$1,428,571	$1,428,571	$1,428,571	$7,005,213
Annual life-cycle cost	$2,646,401	$2,928,571	$2,928,571	$2,928,571	$2,928,571	$14,360,686

Table 11.14
Transportation Management Cost Estimate

Transportation Management					
Element	Smart City Transportation Central	Traffic Management Central	Traffic Signal Central	Transit Management Central	Totals
Scale factor (capital per unit)	$10,000	$200,000	$10,000	$20,000	
Units	1,538	51	1,115	539	
Total capital investment required	$15,380,000	$10,200,000	$11,150,000	$10,780,000	$47,510,000
Annual operating costs	$2,307,000	$1,530,000	$1,672,500	$1,617,000	$7,126,500
Assumed design life in years	7	7	7	7	
Annual capital investment required	$2,197,143	$1,457,143	$1,592,857	$1,540,000	$6,787,143
Annual life-cycle cost	$4,504,143	$2,987,143	$3,265,357	$3,157,000	$13,913,643

Table 11.15
Traveler Information Cost Estimate

Traveler Information				
Element	Traveler Information Central	Smart Phone Apps Subsystem	Movement Analytics Subsystem	Totals
Scale factor (capital per unit)	$40	$2	$2	
Units	1,000,000	1,000,000	1,000,000	
Total capital investment required	$40,000,000	$2,000,000	$2,000,000	$44,000,000
Annual operating costs	$6,000,000	$300,000	$300,000	$6,600,000
Assumed design life in years	7	1	1	
Annual capital investment required	$5,714,286	$2,000,000	$2,000,000	$9,714,286
Annual life-cycle cost	$11,714,286	$2,300,000	$2,300,000	$16,314,286

sign life of seven years has been assumed for all elements. The number of urban delivery vehicles is assumed to be 10% of the total number of private cars.

For the connected, involved citizens and visitors system, it is assumed that the cost of the central hardware and software is proportional to the number of

Table 11.16
Urban Automation Cost Estimate

Urban Automation						
Element	Urban Automation Central	Urban Delivery Truck In-vehicle System	Rental car and taxi in-vehicle system	Bus in-vehicle system	Private car in-vehicle system	Totals
Scale factor (capital per unit)	$15	$1,000	$1,000	$1,000	$1,000	
Units	903,649	80,799	14,321	539	807,990	903,649
Total capital investment required	$13,554,735	$80,799,000	$14,321,028	$539,000	$807,990,000	$917,203,764
Annual operating costs	$2,033,210	$12,119,850	$2,148,154	$80,850	$121,198,500	$137,580,565
Assumed design life in years	7	7	7	8	7	
Annual capital investment required	$1,936,391	$11,542,714	$2,045,861	$67,375	$115,427,143	$131,019,484
Annual life-cycle cost	$3,969,601	$23,662,564	$4,194,015	$148,225	$236,625,643	$268,600,049

Table 11.17
Urban Delivery and Logistics Cost Estimate

Urban Delivery And Logistics			
Element	Urban Delivery And Logistics Central	Urban Delivery Truck In-vehicle System	Total
Scale factor (capital per unit)	$150	$500	
Units	80,799	80,799	
Total capital investment required	$12,119,850	$40,399,500	
Annual operating costs	$1,817,978	$6,059,925	
Assumed design life in years	7	7	
Annual capital investment required	$1,731,407	$5,771,357	
Annual life-cycle cost	$3,549,385	$11,831,282	$15,380,667

citizens and visitors using the system. A scale factor of five dollars per person or smart phone has been assumed. The connected citizen and visitor application is assumed to cost $10 USD per smart phone or person to develop, market, and install. A design life of seven years is assumed for the central hardware and

Table 11.18
User-Focused Mobility Cost Estimate

User-Focused Mobility					
Element	User Focused Mobility Central	User Subsystem	Public Mobility Service Provider Subsystem	Private Mobility Service Provider Subsystem	Totals
Scale factor (capital per unit)	$50	$10	$50	$50	
Units	1,000,000	1,000,000	1,000,000	1,000,000	
Total capital investment required	$50,000,000	$10,000,000	$50,000,000	$50,000,000	$160,000,000
Annual operating costs	$7,500,000	$1,500,000	$7,500,000	$7,500,000	$24,000,000
Assumed design life in years	7	1	7	7	
Annual capital investment required	$7,142,857	$10,000,000	$7,142,857	$7,142,857	$31,428,571
Annual life-cycle cost	$14,642,857	$11,500,000	$14,642,857	$14,642,857	$55,428,571

Table 11.19
Intelligent Sensor–Based Infrastructure Cost Estimate

Intelligent Sensor–Based Infrastructure		
Element	Sensors	Total
Scale factor (capital per unit)	$20,000	
Units	5,900	
Total capital investment required	$118,000,000	$118,000,000
Annual operating costs	$17,700,000	$17,700,000
Assumed design life in years	7	
Annual capital investment required	$16,857,143	$16,857,143
Annual life-cycle cost	$34,557,143	$34,557,143

software, while a design life of one year is assumed for the connected citizen and visitor application. This reflects the disposable nature of smart phone apps and the pace of change regarding the small software applications.

For the integrated payment system, the cost of parking, a smart phone payment application, toll collection, and transit ticketing have been defined separately and then combined into a single integrated payment system cost. For the electronic parking subsystem, it is assumed that the central hardware and

Table 11.20
Low-Cost Efficient, Secure, and Resilient ICT Cost Estimate

Low-Cost Efficient, Secure, And Resilient ICT		
Element	Communications Infrastructure	Total
Scale factor (capital per unit)	$50,000	
Units	5,900	
Total capital investment required	$295,000,000	$295,000,000
Annual operating costs	$44,250,000	$44,250,000
Assumed design life in years	20	
Annual capital investment required	$14,750,000	$14,750,000
Annual life-cycle cost	$59,000,000	$59,000,000

Table 11.21
Smart City Transportation Service Direct Benefit Cost Summary

Smart City Transportation Service Cost Summary	Annual Life-Cycle Cost
Asset and maintenance management	$5,183,571
Connected vehicle	$145,547,227
Connected, involved citizens	$12,964,286
Integrated electronic payment	$36,332,842
Smart grid, roadway electrification and electric vehicle	$144,597,981
Smart land use	$14,360,686
Transportation management	$13,913,643
Traveler information	$16,314,286
Urban automation	$268,600,049
Urban delivery and logistics	$15,380,667
User-focused mobility	$55,428,571
Direct benefit services total	$728,623,810
Cost contingency of 10%	$72,862,381
Direct benefit services total with contingency of 10%	$801,486,191

software will cost approximately $100 USD per parking space. For the smart phone payment application, it is assumed that the cost of software development, marketing and installation will be on the order of $5 USD per person. For the electronic toll collection subsystem, a scale factor of $50 USD per vehicle has been assumed and for electronic transit ticketing and scale factor of $50,000 USD per bus has been assumed. Central hardware and software is assumed to have a design life of seven years, while the smart phone application software is assumed to have a design life of one year.

For the smart grid, roadway electrification, and electric vehicle system, the central, roadside, and in-vehicle subsystem have been costed separately and

Table 11.22
Transportation Governance Cost Estimate

Transportation Governance		
Element	Support and Administration of Transportation Governance as a Proportion of Annual Life-Cycle Cost	Total
Scale factor (capital per unit)	1%	
Annual life-cycle operating costs	$801,486,191	$801,486,191
Total capital investment required	$8,014,861.91	$8,014,862
Annual operating costs	$1,202,229	$1,202,229
Assumed design life in years	7	
Annual capital investment required	$1,144,980	$1,144,980
Annual life-cycle cost	$2,347,210	$2,347,210

Table 11.23
Urban Analytics Cost Estimate

Urban Analytics		
Element	Establishing Data Platform and Conducting Analytics	Total
Total cost of ownership per terabyte per year	$12,000	
Terabytes per year	365	
Assumed design life in years	7	
Annual life-cycle cost	$4,380,000	$4,380,000

then combined into a single cost estimate. The design life of all elements is assumed to be seven years. The total number of electric vehicle charging points has been estimated to be 381. This is based on the current density of gas stations per capita in the United States, proportioned to the population of the smart city. A figure of $10,000 USD per electric vehicle charging point has been allocated to establish the cost of the hardware and software for the smart grid management central subsystem, and it has been assumed that each roadside EV charging point will cost $100,000 USD. The in-vehicle subsystem has been assumed to cost $500 USD in addition to the cost of the vehicle.

It is assumed that the smart land-use system will consist of a central hardware and software platform, linked to separate movement analytics, smart factory, smart education, and smart healthcare subsystems. These subsystems will provide data to the smart land-use central subsystem, and the additional cost of doing so is incorporated as part of the cost. It is likely that the cost of the smart factory, smart education and smart healthcare systems will exceed the assumptions here as the functionality and capability of these systems will be designed

Table 11.24
Smart City Transportation Service Cost Summary

Smart City Transportation Service Cost Summary	Annual Life-Cycle Cost
Asset and maintenance management	$5,183,571
Connected vehicle	$145,547,227
Connected, involved citizens	$12,964,286
Integrated electronic payment	$36,332,842
Smart grid, roadway electrification and electric vehicle	$144,597,981
Smart land use	$14,360,686
Transportation management	$13,913,643
Traveler information	$16,314,286
Urban automation	$268,600,049
Urban delivery and logistics	$15,380,667
User-focused mobility	$55,428,571
Direct benefit services total	$728,623,810
Cost contingency of 10%	$72,862,381
Direct benefit services total with contingency of 10%	$801,486,191
Intelligent sensor–based infrastructure	$34,557,143
Low-cost efficient, secure, and resilient ICT	$59,000,000
Urban analytics	$4,380,000
Strategic business models and partnering	$2,347,210
Transportation governance	$2,347,210
Indirect benefit services or enablers total	$102,631,562
Cost contingency of 10%	$10,263,156
Indirect benefit services or enablers total with contingency of 10%	$112,894,718
Cost of enablers as a proportion of total	14%

for specific uses within factories, educational establishments, and healthcare facilities. The cost of the hardware and software for smart land-use planning central is related to the total number of vehicles within the smart city, assuming that each vehicle will add $10 USD to the cost of the central hardware and software, as a scale factor. The movement analytics subsystem hardware and software has been costed in the same manner, using $10 USD per person using the system, as the scale factor. For smart factories, smart education, and smart healthcare, it is assumed that every facility adds $50,000 USD to the cost of the overall system, and this is used as a scale factor.

The transportation management system is comprised of a smart city transportation center that encompasses all modes, a traffic management center, a traffic signal management center, and a transit management center. It has been assumed that none of these centers currently exist, and the cost of establishing them has been built-in to the overall cost. The smart city transportation management center and the traffic signal control center are assumed to cost $10,000 USD for every mile of road managed. The transit management center it is assumed to cost $20,000 USD per bus managed. The cost of the traffic management center, which mainly addresses freeway incident management is assumed to be $200,000 USD for every mile managed. This is a higher figure due to the smaller number of road miles managed. These assumptions yield a capital cost of around $10–11 million USD for each center, which aligns with published information on the cost of transportation management centers [19].

The traveler information system is assumed to be comprised of a traveler information central hardware and software platform, a smart phone application, and a movement analytics subsystem. The movement analytics subsystem is assumed to be a marginal additional cost to the cost of operating the smart phone and has been assumed to be the same cost as a smart phone app at two dollars for each person addressed. The cost for the traveler information central platform is calculated using a scale factor of $40 USD for each person addressed. The cost of the smart phone application development, marketing and installations is assumed to be $2 USD per person addressed and the movement analytics subsystem is assumed to be to be $2 USD per person addressed. A design life of seven years has been assumed for the central hardware and software. Both the smart phone application software and the movement analytics subsystem have an assumed design life of one year.

The urban automation system is assumed to comprise an urban automation central hardware and software platform and urban delivery truck, rental car and taxi, and private car in-vehicle subsystems. The urban automation central hardware and software platform adopts a scale factor of $15 USD per vehicle across the total number of vehicles in a smart city. This includes urban delivery trucks, rental cars and taxis, and private cars. The cost of in-vehicle systems for all vehicle types is assumed to be $1,000 USD in addition to the cost of the vehicle. The design life of all system elements is assumed to be seven years.

The urban delivery and logistics system is assumed to be composed of an urban delivery and logistics central hardware and software platform and an urban delivery truck in-vehicle subsystem. The cost of the central hardware and software is estimated using a scale factor of $150 USD for every vehicle managed, and the cost of the in-vehicle system is assumed to be $500 USD in additional cost over and above the cost of the vehicle. Design life of all elements is assumed to be seven years.

The cost of the user-focused mobility system is assumed to be directly proportional to the population addressed. The cost of the user-focused mobility central hardware and software is estimated using a scale factor of $50 USD per capita. The cost of the user subsystem is estimated using a scale factor of $10 USD per capita The public mobility service provider subsystem and the private mobility per service provider subsystem are estimated using a scale factor of $50

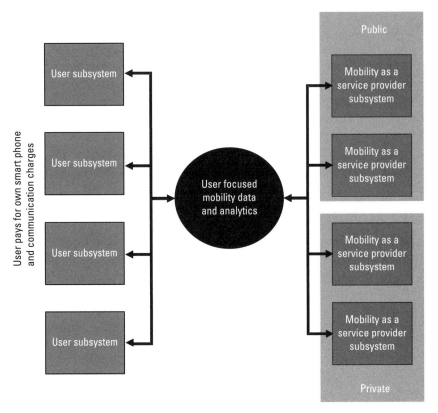

Figure 11.13   Assumed configuration for user-focused mobility system.

USD per capita. The design life for all central hardware and software is assumed to be seven years; the design life of the user subsystem smart phone application is assumed to be one year.

The approach taken toward cost estimation for intelligent sensor–based infrastructure assumes a scale factor of $20,000 USD per sensor connected to the infrastructure. This includes the cost of the sensor and the communication infrastructure required to connect the sensor to a central back office. Sensor count is assumed to be the total number of sensors on all road types and bus routes as summarized in Table 11.5. The design life of seven years has been assumed to account for changes in technology and the technology refresh at year seven.

A scale factor of $50,000 USD per mile of highway and bus routes has been used to estimate the cost of low-cost, efficient, secure, and resilient ICT. This is higher than the figure used for intelligent sensor–based infrastructure as it is assumed that a larger number of devices will be connected. It is assumed that the system cost will predominantly consist of telecommunications hardware such as conduit and fiber optics and that these are less susceptible to technology changes. Therefore, a design life of 20 years has been assumed.

Strategic business models and partnering, like intelligent sensor–based infrastructure, is assumed to be an enabling service that does not deliver direct value in terms of benefits. These services are essential to the delivery of the other services but do not provide direct tangible benefits. It has been assumed that support and administration relating to strategic business models and partnering will be a proportion of the total cost of those services that provide direct benefits. A best estimate of 1% of total cost is been assumed. Table 11.17 summarizes the cost of the direct benefit services. This includes the contingency of 10% to address the high-level nature of the cost estimates.

Transportation governance is treated in a similar manner to strategic business models and partnering. It is assumed that administration and support for transportation governance will represent 1% of the total cost of the direct benefit services.

For urban analytics, it is assumed that the total cost of ownership of each terabyte of data to be stored and managed will be $12,000 USD per year [8]. This includes a complete menu of cost items including hardware depreciation, software purchase or depreciation, maintenance, storage management labor, power consumption, system monitoring, and the cost of procurement [8]. The number of terabytes to be managed by the system is assumed to be 1 TB per day. This aligns with the San Diego Association of Governments experience with its Integrated Corridor Management Project [20]. A design life of seven years is assumed, aligning with the same assumption for other back-office hardware and software in this chapter.

11.9 Smart City Transportation Service Cost Summary

Table 11.20 is based on the calculations illustrated in Section 11.8 and captures the annual life-cycle cost in terms of amortized capital investment and annual operating cost for each of the smart city transportation services. As discussed previously, most of the services deliver direct benefits, while some of the services (e.g., strategic business models and partnering and transportation governance) act as enablers to the direct benefit services. These services do not deliver direct benefits. The enabling services have been treated separately at the bottom of Table 11.20. A 10% contingency has been added to the enabler cost estimates in addition to the 10% contingency added to the direct benefit services earlier.

11.10 Estimated Benefits for Smart City Transportation Services

Estimated benefits for smart city transportation services that provide direct benefit were developed as shown in Tables 11.25–11.35, representing each smart

Table 11.25

Asset and Maintenance Management Benefit Estimate

Asset and Maintenance Management		
	Efficiency Benefit	Total
Annual life-cycle cost	$5,183,571	$5,183,571
Assumed reduction	10%	
Annual life-cycle savings	$518,357	$518,357

Table 11.26

Connected Vehicle Benefit Estimate

Connected Vehicle	Safety Benefit	Connected Vehicle	Efficiency benefit	Total
Fatalities per 100,000 population per year	10.69	Cost of emissions per vehicle mile traveled private car	$0.011	
Injured persons per 100,000 population per year	752	Cost of emissions per vehicle mile traveled urban truck	$0.026	
Fatalities	107	Total cost of emissions private car	$91,606,284	
Injuries	7,520	Total cost of emissions urban truck	$28,968,056.00	
Average cost of fatality	$1,400,000	Operating cost per mile private car	$0.580	
Average cost of injury accident	$1,000,000	Operating costs per mile urban truck	$0.708	
Cost of fatalities	$149,660,000	Total Operating cost of Private car	$4,830,149,520	
Cost of injuries	$7,520,000,000	Total operating cost of urban truck	$788,822,448	
Total cost of accidents	$7,669,660,000	Total cost of operation and emissions	$5,739,546,308	
Assumed reduction	5%	Assumed reduction	5%	
Totals	$383,483,000		$286,977,315	$670,460,315

city transportation service. A commentary on the calculation is described after each table.

The benefits for the asset and maintenance management service are derived on an assumption that the use of the service will result in a 10% cost reduction for asset and maintenance management life-cycle costs. This generates annual life-cycle savings of $518,443 USD.

For the connected vehicle service, annual benefits are assumed to be derived from two sources, a reduction in the cost of fatal and injury accidents and

Table 11.27
Connected, Involved Citizens and Visitors Benefit Estimate

Connected, Involved Citizens	
Efficiency Benefit	
Time spent traveling per person per day in hours	1.325
Time spent traveling per person per year	344.5
Cost of time spent traveling per year	$344,500,000
Average value of travel time savings	$13
Total cost of time spent traveling per year	$4,306,250,000
Assumed reduction	2%
Annual benefits	$86,125,000

Table 11.28
Integrated Electronic Payment Benefit Estimate

Integrated Electronic Payment	
Efficiency Benefit	
Time spent traveling per person per day in hours	1.325
Time spent traveling per person per year in hours	344.5
Total time spent traveling per year	344,500,000
Average value of travel time savings	$12.50
Total cost of time spent traveling per year	$4,306,250,000
Assumed reduction	10%
Annual benefits	$430,625,000

a reduction in the cost of operation and emissions for private cars and trucks. These benefits are derived from better route guidance and decision-quality information delivered into the vehicle and using connected vehicle technology to implement strategies such as usage-based insurance to achieve an improvement in safety. A 5% reduction in both types of accidents and emissions and operating cost for vehicles in the smart city.

Benefits for the connected, involved citizens and visitors service have been estimated based on a 2% reduction in the time spent traveling. It is assumed that better information regarding transportation choices can be provided to citizens and visitors through the service and that this will lead to a reduction in the time spent traveling. There could also be additional benefits in the form of a reduction in the cost of travel, but this has not been included in the benefits calculation since the cost of travel depends on numerous factors including the level of tolls, ticket prices, and parking fees in the smart city.

Integrated electronic payment services benefits are derived on an assumption that 10% of the time spent traveling per person every day can be saved

Table 11.29
Smart Grid, Roadway Electrification, and Electric Vehicle Benefit Estimate

Smart Grid, Roadway Electrification, And Electric Vehicle	Efficiency benefit
Cost of emissions per vehicle mile traveled, private car	$0.011
Cost of emissions per vehicle mile traveled, urban truck	$0.026
Total cost of emissions, private car	$91,606,284
Total cost of emissions, urban truck	$28,968,056
Cost of operation per mile, private car	$0.580
Cost of operation per mile, urban truck	$0.708
Total cost of operation, Private car	$4,830,149,520
Total cost of operation, urban truck	$788,822,448
Total cost of operation and emissions	$5,739,546,308
Assumed reduction in emissions	100%
Assumed reduction in operating cost	76%
Annual benefits emissions	$120,574,340.00
Annual benefits operating costs	$4,262,668,390
Annual life-cycle benefits	$4,383,242,730

Table 11.30
Smart Land-Use Benefit Estimate

Smart Land Use	Efficiency Benefit
Cost of emissions per vehicle mile traveled, private car	$0.011
Cost of emissions per vehicle mile traveled, urban truck	$0.026
Total cost of emissions, private car	$91,606,284
Total cost of emissions, urban truck	$28,968,056
Cost of operation per mile, private car	$0.580
Cost of operation per mile urban truck	$0.708
Total cost of operation, private car	$4,830,149,520
Total cost of operation urban truck	$788,822,448
Total cost of operation and emissions	$5,739,546,308
Assumed reduction	2%
Annual benefits	$114,790,926

through the more efficient payment for transportation, which will speed boarding and improve connectivity between modes and interchange points between transportation routes and services.

The smart grid, roadway electrification, and electric vehicle service is assumed to result in a reduction in emissions and operating cost for private

Table 11.31

Transportation Management Benefit Estimate

Transportation Management	Safety Benefit	Efficiency benefit		Efficiency benefit		Total
Fatalities per 100,000 population per year	10.69	Time spent traveling per person per day in hours	1.325	Cost of emissions per vehicle mile traveled private car	$0.011	
Injured persons per 100,000 population per year	752	Time spent traveling per person per year	344.5	Cost of emissions per vehicle mile traveled urban truck	$0.026	
Fatalities	107	Total time spent traveling per year	344,500,000	Total cost of emissions private car	$91,606,284	
Injuries	7,520	Average value of travel time savings	$12.50	Total cost of emissions urban truck	$28,968,056	
Average cost of fatality	$1,400,000	Total cost of time spent traveling per year	$4,306,250,000	Cost of operation per mile private car	$0.580	
Average cost of injury accident	$1,000,000			Cost of operation per mile urban truck	$0.708	
Cost of fatalities	$149,660,000			Total cost of operation Private car	$4,830,149,520	
Cost of injuries	$7,520,000,000			Total cost of operation urban truck	$788,822,448	
Total cost of accidents	$7,669,660,000			Total cost of operation and emissions	$5,739,546,308	
Assumed reduction	2%	Assumed reduction	2%	Assumed reduction	2%	
Annual benefits	$153,393,200	Annual benefits	$86,125,000	Annual benefits	$114,790,926	$354,309,126

vehicles and urban trucks. A 100% reduction in emissions is assumed as electric vehicles will produce zero omissions. It could be argued that some omissions will simply be transferred from the roadside to the power plant, but it is a reasonable assumption that centralizing these omissions will enable effective and efficient treatment, compared to the distributed omissions from internal combustion engine vehicles. An emissions reduction of 76% for electronic vehicles

Table 11.32

Traveler Information Benefit Estimate

Traveler Information	Efficiency Benefit
Time spent traveling per person per day in hours	1.325
Time spent traveling per person per year	344.5
Total time spent traveling per year	344,500,000
Average value of travel time savings	12.5
Total cost of time spent traveling per year	$4,306,250,000
Assumed reduction	2%
Annual benefits	$86,125,000

Table 11.33

Urban Automation Benefit Estimate

Urban Automation				
Safety Benefit		**Efficiency Benefit**		**Total**
Fatalities per 100,000 population per year	10.69	Cost of emissions per vehicle mile traveled, private car	$0.011	
Injured persons per 100,000 population per year	752	Cost of emissions per vehicle mile traveled, urban truck	$0.026	
Fatalities	107	Total cost of emissions private car	$91,606,284	
Injuries	7,520	Total cost of emissions, urban truck	$28,968,056	
Average cost of fatality	$1,400,000	Cost of operation per mile, private car	$0.580	
Average cost of injury accident	$1,000,000	Cost of operation per mile, urban truck	$0.708	
Cost of fatalities	$149,660,000	Total cost of operation, private car	$4,830,149,520	
Cost of injuries	$7,520,000,000	Total cost of operation urban truck	$788,822,448	
Total cost of accidents	$7,669,660,000	Total cost of operation and emissions	$5,739,546,308	
Assumed reduction	45%	Assumed reduction	10%	
Annual benefits	$3,451,347,000	Annual benefits	$573,954,631	$4,025,301,631

was determined by assuming an average operating cost of $0.14 USD per mile [17] for an electric vehicle compared with $0.58 USD per mile for an internal combustion part vehicle. It is assumed that this cost reduction applies to both private cars and urban trucks.

The smart land-use service is assumed to produce efficiency benefits related to the cost of emissions and cost of operations for private cars and urban

Table 11.34
Urban Delivery and Logistics Benefit Estimate

Urban Delivery and Logistics	Efficiency Benefit		Efficiency Benefit	Total
Cost of urban delivery and logistics	$10,000,000	Cost of emissions per vehicle mile traveled, private car	$0.011	
		Cost of emissions per vehicle mile traveled, urban truck	$0.026	
		Total cost of emissions, private car	$91,606,284	
		Total cost of emissions, urban truck	$28,968,056	
		Cost of operation per mile, private car	$0.580	
		Cost of operation per mile, urban truck	$0.708	
		Total cost of operation, private car	$4,830,149,520	
		Total cost of operation, urban truck	$788,822,448	
		Total cost of operation and emissions	$5,739,546,308	
Assumed reduction	5%	Assumed reduction	5%	
Annual benefits	$500,000	Annual benefits	$286,977,315	$287,477,315

trucks. It is assumed that the service will result in optimized land use and better forecasts of transportation demand for each land-use type and that these together will result in an efficiency gain of approximately 2%.

The transportation management service is assumed to generate benefits in terms of accident reduction, time spent traveling, and the cost of emissions and operations. A 2% reduction in each of these costs has been assumed.

Traveler information services are assumed to generate benefits associated with a reduction in time spent traveling per person per day. Due to better traveler information and decision-quality information provided to travelers at the right time in the right place, it is assumed that travelers will benefit by a 2% reduction in time spent traveling overall.

For the urban automation service, it is assumed that a 45% reduction in accidents will be experienced and the cost of emissions and vehicle operations for both private cars and urban trucks will decrease by 10%. The higher value of a 45% reduction in accidents is assumed based on a published study [21]. This study suggested that automation could lead to a 90% reduction in accidents. However, a midway point of 45% has been assumed to take account

Table 11.35
User-Focused Mobility Benefit Estimate

User-Focused Mobility	Efficiency Benefit		Efficiency Benefit	Total
Cost of emissions per vehicle mile traveled, private car	$0.011	Time spent traveling per person per day in hours	1.325	
Cost of emissions per vehicle mile traveled, urban truck	$0.026	Time spent traveling per person per year	344.5	
Total cost of emissions, private car	$91,606,284	Total time spent traveling per year	344,500,000	
Total cost of emissions, urban truck	$28,968,056	Average value of travel time savings	$12.50	
Cost of operation per mile, private car	$0.580	Total cost of time spent traveling per year	$4,306,250,000	
Cost of operation per mile, urban truck	$0.708			
Total cost of operation, private car	$4,830,149,520			
Total cost of operation, urban truck	$788,822,448			
Total cost of operation and emissions	$5,739,546,308			
Assumed reduction	5%	Assumed reduction	5%	
Annual benefits	$286,977,315	Annual benefits	$215,312,500	$502,289,815

of a gradual transition toward full automation. A 10% reduction in the cost of emissions and vehicle operations has been assumed.

Benefits for the urban delivery and logistics service are generated by a 5% reduction in the cost of urban delivery and a 5% reduction in the cost of emissions and operations for urban vehicles. The cost of urban delivery and logistics is based on an assumed average cost for each urban delivery of $10 USD and an average of 5,000 deliveries per day in the smart city. It has also been assumed that there will be 200 operating days per year, assuming that most of the deliveries will be conducted on weekdays. The cost of emissions and vehicle operation is based on national vehicle miles traveled per year with an assumed split of 88% private vehicles and 12% urban delivery trucks.

For the user-focused mobility service, it is assumed that benefits are derived from a 5% reduction in the cost of emissions and vehicle operations within the smart city and a 5% reduction in the time spent traveling.

11.11 Smart City Transportation Services Cost and Benefits Summary

To summarize the cost and benefits estimation process, Table 11.36 captures the estimated cost and benefits for each of the smart city transportation services. It also shows a benefit cost ratio for each of the direct benefit services. Note once again that five of the smart city transportation services are enablers that do not provide direct benefits and consequently have a cost associated with them but not a benefit and, hence, no benefit-cost ratio.

Table 11.36
Cost and Benefits Estimate Summary

Benefits Summary	Annual Life-Cycle Benefits	Lifecycle Cost	Benefit-Cost Ratio
Asset and maintenance management	$518,357	$5,183,571	0.1
Connected vehicle	$670,460,315	$145,547,227	4.6
Connected, involved citizens	$86,125,000	$12,964,286	6.6
Integrated electronic payment	$430,625,000	$36,332,842	11.9
Smart grid, roadway electrification, and electric vehicle	$4,383,242,730	$144,597,981	30.3
Smart land use	$114,790,926	$14,360,686	8.0
Transportation management	$354,309,126	$13,913,643	25.5
Traveler information	$86,125,000	$16,314,286	5.3
Urban automation	$4,025,301,631	$268,600,049	15.0
Urban delivery and logistics	$287,477,315	$15,380,667	18.7
User-focused mobility	$502,289,815	$55,428,571	9.1
Direct benefit services total	$10,941,265,216	$728,623,810	15.0
Intelligent sensor– based infrastructure		$34,557,143	
Low-cost efficient, secure, and resilient ICT		$59,000,000	
Urban analytics		$4,380,000	
Strategic business models and partnering		$2,347,210	
Transportation governance		$2,347,210	
Indirect benefit enabler services		$102,631,562	
Grand totals	$10,941,265,216	$831,255,372	13.2

11.12   Summary

The subject of benefits and cost for smart city transportation services is a challenging one. Detailed cost estimates are impossible to attain until a detailed design is conducted. This takes into account specific technology choices and the unique attributes of the smart city. This chapter assumes that the definition of a framework or approach toward a smart city transportation service cost-benefit approach will represent a step in the right direction.

It is clear that a more rigorous approach is required in the conduct of before-and-after studies to determine the effects of transportation investments like smart city transportation services. European studies [1] have indicated that in many cases the lack of a formal structured approach to before-and-after studies leads to a scarcity of suitable evaluation material. This points to the need for standards that define the approach to before-and-after studies in smart cities and those that define that require data. It is also obvious that the role of big data and analytics will be crucial in the detailed understanding of the cost and effects of smart city transportation services. It is hoped that the work described in this chapter can act as a foundation for the application of big data and analytics techniques toward the creation of a smart city benefit-cost model.

Another area for further work lies in the assignment of investments between the public and the private sector. This chapter makes no attempt to proportion annual life-cycle costs to specific enterprises or entities. Neither does it attempt to identify the overlap in synergies between the 16 smart city transportation services, as this would be an important role for a detailed cost-benefit model in the future.

The approach defined in the chapter could also form the basis for an incremental planning approach toward the evolution of smart city services over time, over geographic space, and by quality of service. The identification of suitable departure points for smart city initiatives and the exact sequencing of smart city service deployment to achieve optimum effects will very much depend on prior investments and the priority of different policy objectives within the smart city.

The approach identified and described here represents the very beginning of a smart, data-driven, and scientific approach to the determination of investment requirements for smart cities and an understanding of the effects of smart city transportation services. It is hoped that a dialogue can be established between both public- and private-sector enterprises associated with smart city transportation with a view toward refining the approach, aligning the smart city transportation services to available solutions in the marketplace, and making progress toward a detailed cost-benefit model.

References

[1] 2DECIDE project, WP 4.3, deliverable 4.2: reporting validation and testing, version 1.0, European commission, directorate Gen. for mobility and transport, 16 December 2011, http://www.transport-research.info/sites/default/files/project/documents/20120330_130 220_8072_2DECIDE_4_2_Report_on_Validation_and_Testing_final_for_EC_10.pdf, retrieved on March 20, 2017.

[2] https://www.census.gov/popclock/the U.S. Census population clock website, retrieved March 10, 2017.

[3] http://www.nacsonline.com/Research/FactSheets/ScopeofIndustry/Pages/IndustryStoreCount.aspx, the Association for Convenience and Fuel Retailing website, retrieved March 14, 2017.

[4] http://newsroom.aaa.com/2015/04/annual-cost-operate-vehicle-falls-8698-finds-aaa-archive/, AAA website retrieved March 14, 2017.

[5] Transportation Cost and Benefit Analysis II—Air Pollution Cost, Victoria Transport Policy Institute http://www.vtpi.org/tca/tca0510.pdf, retrieved April 10, 2017.

[6] U.S. Department of Transportation, Revised Departmental Guidance on Valuation of Travel Time in Economic Analysis, https://www.transportation.gov/office-policy/transportation-policy/revised-departmental-guidance-valuation-travel-time-economic, retrieved April 10.

[7] Bureau of transportation statistics, total time spent traveling in weekdays and weekends 2003–2014, https://www.rita.dot.gov/bts/publications/passenger_travel_2016/tables/fig2_5_text, retrieved April 10, 2017.

[8] Hitachi Data Systems, White Paper, Storage Economics, Four Principles for Reducing Total Cost of Ownership, April 2015, https://www.hds.com/en-us/pdf/white-paper/four-principles-for-reducing-total-cost-of-ownership.pdf, retrieved on April 9, 2017.

[9] https://www.fhwa.dot.gov/policyinformation/pubs/hf/pl11028/chapter1.cfm, Federal Highway administration policy information website, retrieved on 14 March 2017.

[10] http://www.transitchicago.com/about/facts.aspx , Chicago transit Authority website, retrieved on March 14, 2017.

[11] https://energy.gov/eere/vehicles/fact-841-october-6-2014-vehicles-thousand-people-us-vs-other-world-regions, energy.gov, office of energy efficiency and renewable energy website, retrieved onMarch 14, 2017.

[12] https://www.rita.dot.gov/bts/sites/rita.dot.gov.bts/files/publications/state_transportation_statistics/state_transportation_statistics_2015/chapter-5/table5_3, research and innovative technology administration, retrieved March 11, 2017.

[13] https://www.fhwa.dot.gov/planning/tmip/publications/other_reports/commercial_vehicles_transportation/task4_sect3.cfm, Federal Highway administration planning website, retrieved March 11, 2017 .

[14] https://en.wikipedia.org/wiki/Taxicabs_of_the_United_States,Wikipedia website regarding taxicabs of the United States, retrieved on March 14, 2017.

[15]   http://www.autorentalnews.com/fileviewer/2451.aspx, auto rental used.com, retrieved March 14, 2017.

[16]   Parking infrastructure: energy, emissions, and automobile life-cycle environmental accounting, Mikhail Chester1, Arpad Horvath and Samer Madanat, Published 29 July 2010 • IOP Publishing Ltd , section 2 scenario one, http://iopscience.iop.org/article/10.1088/1748-9326/5/3/034001/meta, retrieved March 12, 2017.

[17]   https://avt.inl.gov/sites/default/files/pdf/fsev/cost.pdf, Idaho National Laboratory, Advanced Vehicle Testing Activity website, retrieved March 14, 2017.

[18]   The Economic and Societal Impact of Motor Vehicle Crashes, 2010, National Highway traffic safety administration, 2015 (revised https://crashstats.nhtsa.dot.gov/Api/Public/ViewPublication/812013), retrieved April 10, 2017.

[19]   San Diego Association of Governments, I 15 Integrated Corridor Management Project, http://www.sandag.org/index.asp?projectid=429&fuseaction=projects.detail , retrieved on 17 March 2017. The specific estimate of 1 TB per day was obtained in an interview with Peter Thompson, Senior Technology Program Analyst, SANDAG.

[20]   Intelligent Transportation Systems Benefits, Cost, Deployment, and Lessons Learned Desk Reference: 2011 update, Intelligent Transportation Systems Joint Program Office, U.S. DOT, http://www.itskr.its.dot.gov/its/benecost.nsf/files/BCLLDepl2011Update/$File/Ben_Cost_Less_Depl_2011%20Update.pdf, page 102, retrieved March 20, 2017.

[21]   Bertoncello, M., D. Wee, *Ten Ways Autonomous Driving Could Redefine the Automotive World*, [Online] McKinsey and Company; New York, June 2015, http://www.mckinsey.com/industries/automotive-and-assembly/our-insights/ten-ways-autonomous-driving-could-redefine-the-automotive-world, retrieved on March 20, 2017.

12

Summary

12.1 Instructional Objectives

There are a number of instructional objectives to be achieved in this chapter. The following is a list of the instructional objectives that have been defined:

- Provide a summary of the essential elements of the book and highlight particularly important points;
- Explain the value to be derived from reading the book;
- Discuss further work for which the book can act as a platform;
- Provide some recommendations on how to act on the information in the book through the delivery of some succinct advice;
- Provide some thoughts on further reading.

12.2 Chapter Word Cloud

As in previous chapters, a word cloud has been generated to provide a general overview of this chapter. Figure 12.1 presents the word cloud for this chapter.

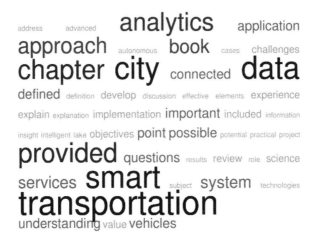

Figure 12.1   Word cloud for Chapter 12.

12.3   Introduction

The primary reason for writing this book at this point in time is to provide a nexus or bridge between transportation in a smart city and data science. There is much interest, thought, and action relating to smart cities at the current time. The subject has, quite rightly, captured the imagination of city planners, mayors, and the public at large. The smart city initiative is acting as a banner that provides a focus point and the convergence catalyst for the application of advanced technologies to urban environments. While the smart city goes beyond transportation into subjects such as smart healthcare, smart places to live and work, and smart utilities, the role of transportation will be inordinately important in the formation and operation of smart cities. It is hoped that this book will add further fuel to the smart cities initiative and provide transportation professionals with some additional insight and understanding to help ensure that smart city transportation investments are made in the most effective manner. One of the early lessons learned in the intelligent transportation system program is the need to identify and harness the common ground between different projects in different applications. Many of the technologies are designed for sharing, and close interaction between the public and private sectors can go a long way to accelerating the deployment programs. This chapter summarizes the preceding chapters and provides an overview of the essential points. It also contains suggestions for further action and further reading.

12.4 Review of Chapter 1

Chapter 1 sets the scene for the remainder of the book, exploring the background and reasoning for writing the book at this time. Chapter 1 also examines the choice of the subject and the timing to explain the value of addressing smart city transportation now. In addition, Chapter 1 explains that the book is aimed at a diverse leadership including transportation professionals, business analysts, automotive professionals, smart city practitioners, and academics and their students. This is a wide audience group, but it is the type of multidisciplinary group likely to work as planners and implementers for smart city transportation. The diversity of the audience group spans multiple disciplines, including big data, connected and autonomous vehicles, smart cities, and data analytics. To round off the book, Chapter 11 has also been included on how to approach benefits and cost estimation for smart city transportation services. The intended effect of the combined information provided in the book is to provide guidance and encouragement to smart city planners and implementers, based on practical experience and expertise drawn from the boundary between smart city transportation and data science.

12.5 Review of Chapter 2

From previous experience in intelligent transportation systems, the introduction of a new subject area often requires that questions are defined before answers are provided. Chapter 2 defines the questions that can be answered by big data and the application of analytics to transportation in the smart city. This serves a twofold purpose. First, it provides a comprehensive list of questions that can form the basis for exploration and discovery. Second, it sets the scene for approaching the answers to the questions and to the content of later chapters. Chapter 2 begins with an explanation of the value of data and why big data techniques and analytics are so important and potentially vital to transportation. The value of data is often underestimated, and it is hoped that the contents of Chapter 2 will help to raise awareness of the value of data for smart city transportation professionals.

Subsequently, Chapter 2 explains why it raises questions rather than providing answers at this stage of the book. Next, Chapter 2 provides a list of questions that can be answered within the smart city transportation realm, structured into safety, efficiency, and user experience categories and listed according to their intended users or audience. The intent is to provide a list of useful questions that can form a starting point for each readership group to develop a customized set of questions pertaining to their specific roles and responsibilities within a smart city. A major component of the approach described in this book revolves around treating transportation as a single system. To this end, Chapter 2 provides four additional questions as a means of assessing the progress we have made toward making transportation behaviors a single system. These relate to clarity of purpose, connectivity, status determination, and adaptability. It is left to the reader to evaluate his or her specific city domain against this four-question checklist.

12.6   Review of Chapter 3

Much use is made of the term big data, and often a clear and agreed-upon definition is missing from the conversation. Chapter 3 addresses the need for such a clear and agreed on definition and provides one that forms the basis for much of the remainder of the book. To begin, Chapter 3 explains how data is measured. While this may be perceived as trivial among some of the readership groups, it could be important to bring everyone to the same page and act as an important element in the bridge that we are building between data science and transportation. A multidimensional explanation of big data is then provided, explaining that big data is certainly about volume, but has other facets as well. These include the variety of the data, variability of the data, data complexity, and veracity of the data. Chapter 3 also explains the difference between real-time analytics and analytics based on archived or static data.

An important element in Chapter 3 is its discussion of how big data and analytics can be perceived as a steppingstone to future automation. There seems to be an incredible focus on smart vehicles and not so much interest in the accompanying smart back office. Accordingly, Chapter 3 explores data management and how the simple ability to have an enterprise-wide view of all the data collected could go a long way toward more efficient and more effective use of the data. To bring the big data discussion to life from a smart city transportation perspective, Chapter 3 concludes by explaining big data within a transportation context, providing specific examples with potentially big data sources within transportation.

12.7   Review of Chapter 4

As discussed earlier, transportation is likely to have an inordinately important role in the smart city. It is also likely that connected and autonomous vehicles will have a disproportionately important role in transportation within a smart city. Chapter 4 explores the definition of the connected vehicle and some of the challenges associated with the introduction of the connected vehicle. In addition, Chapter 4 examines two variations on the connected vehicle theme, with specific emphasis on the telecommunications approach taken to connect vehicles to the roadside or back-office infrastructure. Furthermore, Chapter 4 defines the autonomous vehicle and looks at its associated challenges. In the interest of clarity, In addition, Chapter 4 discusses the differences between connected and autonomous vehicles, and to merge this subject with the smart city information included in Chapter 5, it discusses connected and autonomous vehicles within the context of smart city transportation operation. To conclude, Chapter 4 describes how connected and autonomous vehicles might affect transportation and the automobile manufacturing industry.

12.8 Review of Chapter 5

Smart cities is a central theme of the book. Accordingly, Chapter 5 defines the smart city from a transportation perspective in the form of 16 smart city transportation services. To establish a strong connection between the defined services and the types of needs and objectives found in a smart city, Chapter 5 creates a mapping that relates services to objectives, with examples of several potential smart city objectives from a transportation perspective. Subsequently, Chapter 5 describes how the defined services might play a role in defining departure points and provides a roadmap to move from today's situation to tomorrow's smart city, with a suggested approach to incorporating services within a sketch planning approach to smart city implementation.

As smart city implementations will bring challenges as well as opportunities, Chapter 5 provides an initial list of such challenges along with an explanation on the nature of the challenge. Finally, Chapter 5 outlines some practical lessons gleaned in the implementation of the London congestion charging project. These represent real-world challenges experienced in the application of advanced technologies within an urban transportation environment. To conclude Chapter 5 introduces and explains the concept of a sentient city, which goes beyond the smart city. Sentience involves the ability of a city to sense and react appropriately, acting as a single system and applying intelligence to the results of the sensing. This will result in the application of actionable insight to the development of appropriate responses.

12.9 Review of Chapter 6

As a continuation of the bridge building between data science and transportation, Chapter 6 provides an explanation of data analytics and the wider worlds of industry and commerce. The applicability of this experience to transportation within a smart city is also discussed. In addition, Chapter 6 defines data analytics and compares analytics and KPIs, tackling the essential difference between reporting and analytics—that analytics tend to span multiple data sources and are more likely to consist of ratios and comparisons between different data elements. Analytics also provides a more flexible approach that can answer the questions you have already identified and ones that you will discover in the process of using big data and analytics. The conventional KPI and reporting approach tends to create a rigid framework that has difficulty in accommodating new questions. Chapter 6 also provides an important differentiator in the form of a sporting analogy suggesting that even the best reporting system can only make you a spectator at a sports game. Data analytics are an essential tool to enable you to become the coach and exert an influence over the performance of the team or the organization. Accordingly, Chapter 6 further explores the value of analytics by explaining the key role that they play in uncovering patterns and trends within smart city transportation. In addition, Chapter 6 lists analytics for each of the smart city transportation services previously identified, providing concrete examples of their direct relevance to smart city transportation.

12.10   Review of Chapter 7

Building on the concept of a departure point in steppingstones for smart city evolution, Chapter 7 identifies five possible departure points for a smart city from a transportation perspective. Each of the departure points is examined in terms of the overall nature of the departure point, its suitability as a departure point, and the type of analytics that can be used to support the deployment of the departure point. This approach enables the practical application of smart city transportation analytics to be couched within a possible approach to defining starting points and roadmaps for smart city transportation service evolution. While Chapter 7 does not provide a comprehensive list of possible departure points and only explains a sample of the relevant analytics, it should be sufficient to provide the basis for a customized approach to transportation analytics and service evolution within a smart city.

12.11   Review of Chapter 8

A system engineering tool known as the use case has a disproportionately important role in the bridge between transportation and data science. For transportation professionals within a smart city, the use case captures the problems to be addressed, the analytics to be applied, the data required, and the benefits that will be achieved. From a data science point of view this is essential communication of the customer needs, issues, problems, and objectives, that are vital in guiding the development of big data and analytics approaches. Accordingly, Chapter 8 defines the term use case and provides some examples of smart city transportation use cases that can be used as a model for a customized smart city approach, explaining how the use cases can be implemented in practice, with respect to smart city transportation.

12.12 Review of Chapter 9

Establishment and operation of a smart city transportation data lake is an important consideration if the power and value of analytics is to be unleashed. Current approaches to this include the development of data ingestion platforms that can bring data together from multiple sources, assembling it and then distributing it to many parties. The data lake concept incorporates this kind of smart data switch approach but adds the extremely valuable element of analytics capability and analytic sharing. The creation of a smart city transportation data lake has the potential to provide substantial inputs into decision-making, including performance management and service delivery planning. It also has the capability to act as a strong cohesive force between the separate elements of a smart city implementation by providing a common thread for data and analytic sharing. Experience has shown that simply making the data available across a smart city is not likely to guarantee effective use of the data. Extending the sharing approach to include the analytics that can be derived from the data is likely to provide additional stimulus for good use of data and analytics. Just as we have derived lessons and experiences from prior implementations of intelligent transportation systems, there are some great lessons and experiences to be learned from the creation of data lakes in enterprises beyond transportation. Before proposing an approach to the establishment and operation of a smart city data transportation lake, Chapter 9 provides an overview of the challenges that can be involved in the creation of a smart city transportation data lake. This emphasizes the need for the incorporation of lessons learned, practical experiences, and the adoption of practical, robust approach to the creation of the smart city transportation data lake. To emphasize the value of the proposed approach, Chapter 9 also includes a mapping of the challenges to the essential elements of the proposed approach, explaining the value that can be delivered through the creation and operation of a smart city transportation data lake. Chapter 9 concludes with an exposition of how analytics and the data lake can be used to guide the structure of a smart city transportation organization and provide the foundation for a successful enterprise.

12.13 Review of Chapter 10

To illustrate the value of big data and analytics within a smart city transportation initiative, Chapter 10 defines a range of concepts that relate the application of big data and data analytics techniques to smart city transportation. These cover the spectrum of smart city transportation applications, including freeway speed variability analysis, smart city accessibility analysis, an evaluation of the return an investment provided to the driver in return for toll payment, arterial performance management, and decision support for bus acquisition. While some of the examples are further down the implementation pattern others, Chapter 10 provides clearly explains the practical application of analytics to a range of transportation subjects relevant to a smart city.

12.14   Review of Chapter 11

Chapter 11 explains the conceptual framework for estimating benefits and costs associated with smart city transportation services. While it is not possible to provide detailed cost estimates without detailed design and technology choices, Chapter 11 develops and delivers a range of configuration assumptions as a means of supporting a sketch-planning approach to cost-benefit analysis. To provide some realistic results, it was necessary to assume a model smart city and to develop appropriate parameters that explain the nature and characteristics of the city. While these are as realistic as possible, based on research, the overall cost-benefit estimation approach provides orders of magnitude results, rather than detailed assessments. Chapter 11 also illustrates a structured approach to cost and benefit estimation for smart city transportation services and explains how this might be used as a model for planning and high-level screening evaluation for smart city transportation services. It is hoped that the content of Chapter 11 will form the basis for further work that will result in a more detailed cost-benefit model for smart city transportation services.

12.15   Advice for Smart City Transportation Professionals

To summarize the information and insight covered in this book, 24 pieces of advice are now presented. These suggestions, listed as follows, represent a succinct summary of actions readers should consider once they understand the contents of the book.

1. Smart city transportation is a wonderful opportunity to accelerate the application of advanced technologies for the benefits of city dwellers and visitors. The transportation profession should grasp this opportunity with both hands and harness the new power of data science.

2. Building a bridge between transportation and data science is not sufficient; it is necessary for both parties to begin walking across the bridge and meet somewhere in the middle.

3. The opportunities in big data and analytics are matched with potential challenges. Access to data and analytics power could mean that people outside of transportation and outside of one's enterprise are better informed about smart city transportation planning and operations than the professionals involved in delivery.

4. There is an ongoing challenge associated with working across disciplinary boundaries. As in the case of intelligent transportation systems, smart city transportation—if it is to be done effectively and efficiently—will require the mobilization of multidisciplinary teams who understand the objectives and have, at the least, an awareness of one another's specializations.

5. Defining the questions is a good starting point, because in a new subject area the definition of the questions can help to clarify goals and objectives. It is also difficult to get answers if you do not know what questions to ask.

6. An awareness of big data is fundamental to the successful application of these techniques within a smart city transportation environment. Other areas of industry and commerce may be ahead of transportation in the adoption of these techniques, but the application of lessons learned and practical experiences from these other areas can accelerate transportation and deliver spectacular results. They also say in Silicon Valley that, "The early bird catches the worm, but the second mouse gets the cheese." Being just behind the leaders and early adopters could be excellent positioning to take advantage of lessons learned and prior experience.

7. We have an excellent opportunity to adapt and adopt big data and analytics techniques from other industries and apply them in smart city transportation.

8. There has been a sea change in data science approaches. While previously we were inclined to fragment and partition data to manage it effectively and efficiently, it is now possible to consolidate data and create enterprise-wide views across all data. This enables the maximum probability of data elements combining to create new insight and understanding. Consider the new possibility of a consolidated data repository or data lake for smart city transportation.

9. Big data and analytics has the potential to address the spectrum of smart city transportation activities from planning through design and project delivery to operations and maintenance. It is important that a cross-section of transportation professionals become aware of new capabilities.

10. Develop a detailed understanding of the value of data and the new economics involved in retaining data. It may not be necessary to make an early judgment on the value of every data element as this would preclude later combination to deliver new insights and understanding.

11. This also involves the development of a new perception regarding the current cost of data storage and data management in the age of Hadoop and other big data storage approaches.

12. Develop a smart city transportation data analytics strategy employing use cases and pilot projects to develop a true understanding of objectives and possibilities. In the same vein, use new insight into the possibilities of analytics to drive data collection and data acquisition based on an understanding of the purpose to which the data will be put.

13. Be prepared to partner as new possibilities exist to share cost and to harness private-sector motivation to public sector objectives. We may have little choice but to adopt partnership, advocacy, and influence strategies as a potential private-sector role in transportation service delivery emerges and grows.

14. Be aware of the possibilities of connected and autonomous vehicles. Probe data emanating from connected vehicles has the possibility to revolutionize data capture for transportation needs within a smart city. Similarly, the autonomous vehicle has the potential to address many smart city transportation issues including congestion, parking, and last-mile transportation service delivery. Also, be aware of the potential challenges associated with autonomous vehicles that could lead to disruption in the current transportation service delivery model, unless we anticipate and incorporate these new technologies into our smart city transportation approaches.

15. Adopt an objectives-driven approach to the definition of smart cities initiatives, understanding the relative importance of transportation within the larger smart city technology framework.

16. Be aware of the new possibilities for scientific results-driven investment programs, based on insights and understandings derived from big data and transportation data analytics.

17. Be prepared to understand and take advantage of the ability for one project to support another. While each project will deliver indepen-

dent results, the best results will always be achieved when there is a coordination and harmony between projects and initiatives.

18. There is an emerging possibility to go beyond even the smart city to the sentient city where sensing has been optimized and the level of intelligence is such that appropriate strategies can always be defined and implemented. Big data and analytics can also play a key role in planning and operating transportation as a single system with clarity of purpose, connectivity, adaptability, and effective status determination at any given time.

19. Be prepared to learn lessons and take advantage of practical experience gained from other industries and from other smart city transportation initiatives around the globe.

20. Start working on data definitions for subjective transportation terms. Prepare the way for analytics to be applied to large-scale smart city transportation data sets in order to identify new trends and patterns. Also be aware of the significant differences between reporting and analytics and the need for a deep and wide data repository to get the most from the analytics.

21. Be open to the possibility of new scientific approaches to transportation planning, traffic engineering, and work program definition.

22. Adopt a structured approach to the evolution of smart cities, taking full account of the ability for one project to support another.

23. Develop an awareness of the new possibilities and capabilities of data science and how these advances can be adapted to smart city transportation.

24. Begin the development of a structured cost-benefit analysis framework for smart city transportation services, and make that the basis for a detailed cost-benefit model based on detailed design of proposed implementations.

12.16 Conclusion

This book—by necessity—spans a wide area of interest. As noted previously, a major objective of the book is to enable the building of a bridge between smart city transportation and data science. I have had the privilege of a position at the boundary between transportation and data science that has enabled me to accumulate insight and understanding. As expected, this experience of structuring and documenting experiences gained has resulted in new learning and understanding. It is to be hoped that the book will have the same effect on its readers and fulfill its intended role as an effective tool in the application of advanced technologies to smart city transportation.

12.17 Further Reading

If this book has inspired you to seek further information on the subjects covered, consider the following additional reading:

- *Thinking Highways* magazine [1]: This magazine focuses on the application of advanced technologies to transportation but regularly features focuses and themes regarding smart city transportation.
- Smart City Council [2]: The smart city Council is a unique combination of public agencies and private-sector solution providers offering an excellent range of resources for smart city practitioners.
- Previous books by the Author: *ITS Architectures* [3] and *Advanced Traveler Information Systems* [4]. Both are now a few years old but contain unchanged fundamental principles on which this book is based.
- The National Intelligent Transportation Systems Architecture website [5]: Here again, the material is a few years old, however it has been maintained and updated and represents a phenomenal resource covering multiple aspects that are relevant to the application of advanced technologies to smart city transportation. This includes technical, organizational, and business model aspects.

References

[1]   *Thinking Highways*, http://thinkinghighways.com/.

[2]   Smart city Council, http://smartcitiescouncil.com/.

[3]   *Intelligent Transportation System Architectures*, https://www.amazon.com/Intelligent-Transportation-System-Architecture-Library/dp/089006525X/ref=sr_1_1?ie=UTF8&qid=1490323442&sr=8-1&keywords=intelligent+transportation+system+architectures

[4]   *Advanced Traveler Information Systems*, https://www.amazon.com/Advanced-Traveler-Information-Systems-McQueen/dp/1580531334/ref=sr_1_1?ie=UTF8&qid=1490323508&sr=8-1&keywords=Bob+McQueen+advanced+traveler+information+systems

[5]   National Intelligent Transportation Systems Architecture website, http://local.iteris.com/itsarch/.

About the Author

Bob McQueen is a creative, technical problem solver with superb verbal and written communication skills, combined with proven technical, business planning, business development and marketing experience. He is highly experienced in advanced transportation technology-related business building, opportunity analysis and relationship building. He has particular skills and experience in the use of advanced data analysis techniques, big data, and performance management systems. He has written more than 40 papers on advanced transportation technology subjects and two books on system architecture and traveler information systems. Bob is currently the CEO of Bob McQueen and Associates, based in Orlando, Florida. The company is dedicated to assisting public agencies and private enterprises to understand the effects of technology investments and harness the full potential of advanced transportation technologies, including big data and analytics techniques.

Index

Accessibility index
    analytics, 215
    concept, 197
    overview, 214–15
    use case, 170
    use of analytics, 219
Adaptability question, 19
Advice for smart city transportation
        professionals, 278–81
Analytics
    accessibility index, 215
    application to transportation, 156–57
    arterial performance management, 216
    data lakes and, 133–34
    data needs associated with, 134
    decision support for bus acquisition,
        217
    defined, 34, 119–20
    discovery, 153
    graph, 160
    integrated payment system application,
        141, 142
    introduction to, 117–18
    KPIs versus, 132
    MaaS application, 144, 145
    machine learning, 160
    movement, 214
    nature of process, 153
    performance management, 151–52
    practical application of, 137
    in practice, 121–22
    search pattern and time series, 160

smart city frameworks, 101
smart city services, 122–31
statistical modeling, 160
structured query language (SQL), 160
summary, 134–35
text, 160
toll return index, 212–14
traffic management, 146, 147
transit management, 149
transportation use cases, 160
urban, 85, 91
value chain, 121
value of, 120–22
Application of analytics
    integrated payment system, 139–41
    introduction to, 138–39
    list of, 139
    MaaS, 141–44
    performance management, 149–52
    summary, 152–54
    traffic management, 144–46
    transit management, 146–49
Application of transportation data analytics
    accessibility index, 197, 198–214, 219
    arterial performance management, 197,
        216, 220
    concepts, 197–98
    decision support for bus acquisition,
        198, 216–17, 220
    freeway speed variability analysis, 197,
        198–214, 217–19
    introduction to, 196

Application of transportation data analytics
(continued)
    toll return index, 197, 212–14, 219
    use of analytics, 217–20
Architectures and standards, smart cities, 86
Archive data
    defined, 35
    sources, 51, 52
Arterial performance management
    analytics, 216
    concept, 197
    overview, 216
    use of analytics, 220
Asset and maintenance management
    benefit estimate, 260
    cost estimate, 248
    cost estimation, 238
    services, 88
    smart city services analytics, 124–25
    use case, 162
Automotive electronics, 56, 57
Automotive industry
    connected and autonomous vehicles
        effect on, 77–79
    direction of, 77
    technology as shaping future, 78
Autonomous vehicles
    challenges, 69–71
    convergence, 76
    data sources, 50, 51
    defined, 67
    on-demand transportation services,
        75–76
    effect on automotive industry, 77–79
    effort and interest in, 153
    getting there in stages, 70
    illustrated, 68, 69
    impact on transportation, 74–75
    introduction to, 56
    regulation, 69–70
    safety gains, 58
    in smart cities, 73–74, 75–77
    solution table, 71
    stages of testing, 69
    summary, 79–80
    summary of differences with connected
        vehicles, 72–73
    See also Connected vehicles

Back office, 3

Benefit estimation
    asset and maintenance management,
        260
    assumptions, 232–35
    connected, involved citizens and
        visitors, 261
    connected vehicles, 260
    integrated electronic payment services,
        261
    methodology, 226
    smart grid, roadway electrification, and
        electric vehicle services, 262
    smart land use, 262
    transportation management, 263
    traveler information, 264
    urban automation, 264
    urban delivery and logistics, 265
    user-focused mobility services, 266
    See also Smart city transportation
        services cost and benefit
        estimation
Big data
    application to transportation, 156–57
    archive data, 51, 52
    capture, 43
    challenges, 42–46
    commercial vehicle operations, 49, 50
    complexity, 40–42
    complexity analysis, 43
    connected and autonomous vehicles,
        50, 51
    connectivity and, 75
    curation, 44
    as data science evolution, 34
    defined, 33
    electronic payment, 48–49
    emergency management, 49–50
    features of, 34
    introduction to, 31–32
    maintenance and construction
        operations, 51, 52
    management, 101
    measurement of, 32–33
    public transportation management, 48,
        49
    search, 44
    sharing, 45
    smart cities, 50–51
    storage, 45–46
    traffic management, 47, 48

transfer, 46
in transportation, 46–51
traveler information, 47, 48
type, 35
value of, 32
variability, 40
variety, 39–40
velocity, 36–39
veracity, 42
volume, 35–36
Bottlenecks
as efficiency-related question, 21
location and duration, 204–5, 207
at major intersections, 205, 206
modified template, 209
speed profile template, 202, 204

Capture, big data, 43
Clarity of purpose question, 19
Cloud-based services
evolution of, 66
list of, 65
Commercial vehicle operations data sources, 49, 50
Complexity
analysis, 43
big data, 40–42
information, growth of, 42
Connected, involved citizens and visitors
benefit estimate, 261
cost estimate, 249
cost estimation, 239–40
services, 88
smart cities, 86
smart city services analytics, 125–26
use case, 163–64
vision element, 86
Connected vehicle probe data use case, 162–63
Connected vehicles
approaches, 58–59
benefit estimate, 260
challenges, 50–55
cloud-based, 59
in connected transportation system, 57
cost estimate, 248
cost estimation, 238
data ownership, 62–63
data sources, 50, 51
defined, 58

driver education, 61–62
effect on automotive industry, 77–79
impact on transportation, 22, 74–75
introduction to, 56
investment focus, 79
security, 60–61
in smart cities, 73–74, 75–77, 83
smart city services analytics, 125
as stepping stone, 78
summary, 79–80
summary of differences with
autonomous vehicles, 72–73
telecommunications approach, 63–67
two-way communication capability, 3
vehicle-to-vehicle communications
between, 73
See also Autonomous vehicles
Connectivity question, 19
Control centers, 102
Cost estimation
asset and maintenance management, 238, 248
assumed configurations for, 237–46
assumptions, 232–35
connected, involved citizens and
visitors, 239–40, 248
connected vehicles, 238, 248
ICT, 254
integrated electronic payment services, 240, 241, 249
intelligent sensor-based infrastructure, 253
methodology, 226
smart grid, roadway electrification, and
electric vehicle services, 240–41, 242, 250
smart land use, 241–42, 243, 250
transportation governance, 255
transportation management, 242–43, 244, 251
traveler information, 243–44, 245, 251
urban analytics, 255
urban automation, 245, 246, 252
urban delivery and logistics, 245, 247, 252
user-focused mobility, 245–46, 253
See also Smart city transportation
services cost and benefit
estimation
Curation, big data, 44

Customer relationship management (CRM), 40
Customer satisfaction and travel response use case, 169

Data
  archive, 35, 51, 52
  complexity, 40
  in data lakes, 15
  discovery, 180
  governance, 184
  to information, 32
  ingestion, 180
  in motion, 35
  needs associated with analytics, 134
  open, 99
  orders of magnitude, 33
  ownership, 62–63
  preparation, 180
  raw, 13–15
  real-time, 35
  at rest, 35
  size growth, 33
  speed, 36–37
  storage, 45–46
  transmission times, 46–51
  value of, 15
  See also Big data
Data collection, 175–76
Data lakes
  analytics and, 133–34
  approach addressing of challenges, 191
  approach methodology, 188
  approach to building, 187–90
  building, 173–93
  challenges, 185–87
  combining of data elements, 184
  cost reduction and, 183
  creation elements, 175
  data and analytics exchange, 181
  data discovery, 180
  data governance and, 184
  data ingestion, 180
  data preparation, 180
  data sources, 178–80
  defined, 15, 177–78
  delivery of insight, 181
  elements of, 178–82
  enterprise-wide data access, 182

  as foundation for large-scale proactive analytics, 183
  full-capability roadmap, 189–90
  functioning of, 178–82
  introduction to, 174–77
  model configuration, 179
  organizing for success, 190–93
  pilot project approach methodology, 189
  pilot subjects identification, 188–89
  as platform for innovation, 184–85
  preparation for, 188
  safety, efficiency, and user experience and, 184
  as stepping stone toward automation, 183
  summary, 193
  support for smart city service delivery, 185
  support for smart city services, 181–82
  unused data value and, 184
  value of, 182–85
Data repositories. See Data lakes
Data science revolution, 4
Decision support for bus acquisition
  analytics, 217
  concept, 198
  overview, 216–17
  use of analytics, 220
Dedicated short-range communications (DSRC)
  defined, 59–60
  evolution of, 66
  infrastructure implementation, 66
  speed and latency of, 63
  supported services, 64
Driver education, 61–62

Economy of scale, 176
Efficiency-related questions, 21–27
Electric fleet management use case, 166–67
Electronic payment data sources, 48–49
Emergency management data sources, 49–50
Emerging Technologies Forum, 77
Enterprise resource planning (ERP), 40
Event-based triggering, 39
Extraction, transformation, and loading (ETL), 43

Freeway speed variability analysis

approach, 200–201
concept, 197
initial approach, 201–8
INRIX, 202, 218
overview, 198–200
revised approach, 209–11
speed variability, 201–8
TMC segments, 202–4, 209–10, 218
traffic turbulence analysis, 209–11
use of analytics, 217–19
VSL project, 198–200
See also Application of transportation data analytics
Freight performance management use case, 171

Graph analytics, 160

Hard drive storage costs, 46

Information and Communications Technology (ICT)
analytics, 127
cost estimate, 254
management use case, 166
services, 89
vision element, 86
Information complexity, 42
INRIX, 202, 218
Insurance companies, 78
Integrated electronic payment services
analytics, 126
benefit estimate, 261
cost estimate, 249
cost estimation, 240, 241
smart cities, 88
Integrated payment system
analytics and application, 141, 142
elements, 139–40
as good departure point, 140–41
Intelligent sensor-based infrastructure
analytics, 126–27
cost estimate, 253
smart cities, 88–89
use case, 165
Internet of things (IoT), 73
Interoperability interface, 101

Jobs, access to, 27

KPIs
analytics versus, 132
list of, 133
See also Analytics

Land use
optimization, 24–26
smart cities, 89
See also Smart land use
Light detection and ranging (LIDAR) sensors, 67, 69
London Congestion Charge project
boundary zone, 109
defined, 108–9
implementation, 109
lessons learned from, 108–12

MaaS
analytics and application, 144, 145
defined, 141–42
elements, 143
as good departure point, 144
service portfolio, 143
use case, 172
Machine learning, 160
Maintenance and construction operations data sources, 51, 52
Mobility hub use case, 167

National ITS program
advanced vehicle safety systems, 26
commercial vehicle operations, 25
defined, 23
electronic payment, 25
emergency management, 25
information management, 26
maintenance and construction management, 26
public transportation management, 24
travel and traffic management, 24

On-demand transportation services, 75–76
Open apps and open data, 99
Organization, this book, 6–11

Partnership management use case, 168
Performance management
analytical, for smart cities, 132–33
analytics and application, 151–52

Performance management (continued)
    arterial, 197, 216, 220
    comprehensive approach to, 149–51
    defined, 149
    as good departure point, 151
    parameters, 120
    process, 150
Performance measurements, 16

Questions
    adaptability, 19
    answers and, 16
    categories of, 16
    clarity of purpose, 19
    connectivity, 19
    defining, 18
    efficiency-related, 21–27
    overview, 15–20
    power of, 18
    safety-related, 20–21
    status, 19
    system determination, 19
    transportation industry, 15–20
    20 big, 17
    user experience-related, 27–29
    what to do with, 29

Raw data, 13–15
Readership groups, this book, 5–6
Real-time data, 35
Reporting to automation pathway, 38

Safety-related questions, 20–21
Search, big data, 44
Search pattern and time series analytics, 160
Security
    connected vehicles, 60–61
    smart city frameworks, 101
Sensor platform, 99
Sentient city, 113–14
Service levels
    for citizens, 23
    defined, 23–24
    transportation customer perception, 28
    for visitors, 23–24
Sharing, big data, 45
Situation room, 102
Smart cities

analytical performance management
    for, 132–33
architectures and standards, 86
asset and maintenance management
    services, 88
big picture definition and, 97
challenges, 104–6
characteristics, 233
communication of value and benefits,
    105–6
connected, involved citizens, 86
connected, involved citizen services, 88
connected and autonomous vehicles,
    73–74, 75–77, 83
connected vehicle services, 88
continuous implementation, 98
cost share on project implementation,
    106
data sources, 50–51
defined, 3, 83
growth in interest, 81–82
implementation plan definition and,
    97–98
Information and Communications
    Technology (ICT), 86, 89
infrastructure technologies, 84
initiatives, 82
integrated electronic payment services,
    88
intelligent, sensor-based infrastructure
    services, 88–89
introduction to, 81–83
investment effects, evaluating, 104
legacy definition and, 96–97
number of devices estimation, 233
objectives, 91–92
objectives and transportation services,
    93–94
opportunities, 106–8
partners and multiple disciplines and,
    104–5
phase 1 implementation, 98
policy objectives definition, 95–96
response strategies development, 108
results-oriented focus, 105
sensing capabilities, 76–77
sentient city, 113–14
smart cities council definition, 87

smart grid, roadway electrification, and electric vehicle services, 85–86, 89
smart land use, 86–88
smart land use optimization, 107
smart land use services, 89
smart service evolution approach, 95
steps toward, 92–98
stovepipe projects and, 105
strategic business models and partnering, 85
strategic business models and partnering services, 89
summary, 114
sunk investments in legacy systems and, 105
transportation effects on environment, 107
transportation governance services, 90
transportation management services, 90
transportation services improvement, 107
travel information services, 90–91
understanding possibilities and, 96
urban analytics, 85
urban analytics services, 91
urban automation, 83
urban automation services, 91
urban delivery and logistics, 85
urban delivery and logistics services, 91
urban-focused mobility services, 91
U.S. DOT vision elements, 85
user-focused mobility services, 84
Smart city frameworks
analytics, 101
citizens participation platform, 99
control centers, 102
event repository and big data management, 101
information infrastructure, 99
interoperability interface, 101
open apps and open data, 99
organizational framework, 103
overview, 98–99
processes, 101
safety and security, 101
SDK external integrators, 101
semantics and ontology, 101
sensor platform, 99
situation room, 102
software applications and services, 102
technology framework, 100
video platform, 99
Smart city initiatives, 152
Smart city services analytics
asset and maintenance management, 124–25
connected vehicles, 125
Information and Communications Technology (ICT), 127
integrated electronic payment, 126
intelligent sensor-based infrastructure, 126–78
list of, 123
overview, 122–24
smart grid, roadway electrification, and electric vehicle services, 127
smart land use, 128
strategic business models and partnering, 182
transportation governance, 128–29
transportation management, 129–30
urban analytics, 130
urban automation, 130–31
urban delivery and logistics, 131
user-focused mobility, 131
Smart city transportation professionals, advice for, 278–81
Smart city transportation services
categorized as direct or indirect benefit, 236
cost estimates for, 246–59
cost summary, 256, 259
defining, 225, 235
direct benefit cost summary, 254
mapped to efficiency objectives, 228–30
mapped to safety objectives, 227
mapped to user experience objectives, 231
Smart city transportation services cost and benefit estimation
assumed configurations for cost estimation, 237–46
assumptions, 232–35
benefit estimation, 259–66
cost estimation, 246–49
introduction to, 224–25
methodology, 226
overview of approach, 225–32
summary, 267

Smart grid, roadway electrification, and
     electric vehicle services
     analytics, 127
     benefit estimate, 262
     cost estimate, 250
     cost estimation, 240–41, 242
     smart cities, 89
     vision element, 85–86
Smart information infrastructure, 99
Smart land use
     benefit estimate, 262
     cost estimate, 250
     cost estimation, 241–42, 243
     optimization, 107
     smart cities, 107
     smart city services analytics, 128
     vision element, 86–88
Software applications and services, 102
Speed variability
     bottleneck location and duration,
          204–5, 207
     bottlenecks at major intersections, 205,
          206
     bottleneck speed profile template, 202,
          204
     peak period weekdays, 202, 203
     two directions for full year, 201
     See also Freeway speed variability
          analysis
Statistical modeling, 160
Status question, 19
Storage, big data, 45–46
Strategic business models and partnering
     analytics, 128
     services, 89
     vision element, 85
Structured query language (SQL), 160

Teradata, Inc., 3
Text analytics, 160
Ticketing strategy and payment channel
     evaluation use case, 164–65
Toll return index
     analytics, 212–14
     concept, 197, 212
     efficiency benefits, 213
     safety benefits, 212
     toll paid, 212
     use of analytics, 219–20

user experience enhancement benefits,
     213–14
Traffic management
     analytics and application, 146, 147
     automation, 153–54
     data sources, 47, 48
     elements, 144–45
     as good departure point, 146
Traffic Message Channel (TMC) segments,
     202–4, 209–10, 218
Traffic turbulence analysis, 209–11
Transfer, big data, 46
Transit management
     activities, 146–48
     analytics and application, 149
     fleet management, 146
     as good departure point, 148
     passenger information, 148
     ticketing, 148
Transportation
     asset performance, 22
     big data in, 46–51
     connected and autonomous vehicle
          impact on, 74–75
     current and future demand, 27
     customer payment, 23
     delivery activities, 41
     effects on environment, 107
     expenditures, 22
     infrastructure, inflexibility of, 19
     infrastructure and operations data, 3
     land use and, 24–26
     performance measurement, 16
     project concept and, 18
     raw data and, 13–15
     supply and demand, 107–8
     user experience enhancement, 27–28
Transportation customers
     best use of transportation system, 29
     service level perception, 28
     value received by, 28
Transportation governance
     analytics, 128–29
     cost estimate, 255
     services, 90
     use case, 168–69
Transportation management
     analytics, 129
     benefit estimate, 263
     cost estimate, 251

cost estimation, 242–43, 244
data sources, 48, 49
services, 90
Transportation system
bottlenecks and slowdowns in, 21
connected and autonomous vehicle
impact on, 22
efficiency optimization, 22
maximizing safety, 20–21
mobility as a service impact on, 28
safety improvement effects, 21
service deficiencies, 26–27
traveler use of, 28–29
Transportation systems management
and operations (TSM&O), 51–52
Transportation use cases
accessibility index, 170
analytics to be used, 160
asset and maintenance management,
162
business benefits, 159
challenges, 159
connected, involved citizens, 163–64
connected vehicle probe data, 162–63
customer satisfaction and travel
response, 169
defined, 157–58
electric fleet management, 166–67
expected outcome of analyses, 158
format, 158–60
freight performance management, 171
ICT management, 166
intelligent sensor-based infrastructure,
165
introduction to, 156–57
list of, 159
MaaS, 172
mobility hub, 167
objectives, 158
partnership management, 168
smart city examples, 158–61
source data examples, 158–59
success criteria, 158
summary, 160–61
ticketing strategy and payment channel
evaluation, 164–65
transportation governance system,
168–69
travel value analysis, 170
urban automation analysis, 171

variable tolling, 164
Traveler information
benefit estimate, 264
cost estimate, 251
cost estimation, 243–44, 245
data sources, 47, 48
services, 90–91
Travel value analysis use case, 170

Urban analytics
cost estimate, 255
services, 91
smart city services analytics, 130
vision element, 85
Urban automation
analysis use case, 171
analytics, 130–31
benefit estimate, 264
cost estimate, 252
cost estimation, 245, 246
services, 91
vision element, 83
Urban delivery and logistics
analytics, 131
benefit estimate, 265
cost estimate, 252
cost estimation, 245, 247
services, 91
vision element, 85
Urban-focused mobility services, 91
U.S. DOT smart city vision elements, 85
Use cases
accessibility index, 170
asset and maintenance management,
162
connected, involved citizens, 163–64
connected vehicle probe data, 162–63
customer satisfaction and travel
response, 169
defined, 158
electric fleet management, 166–67
format, 158–60
freight performance management, 171
ICT management, 166
intelligent sensor-based infrastructure,
165
list of, 159
MaaS, 172
mobility hub, 167
partnership management, 168

Use cases (continued)
    purpose of, 157
    smart city transportation examples,
        158–61
    summary, 160–61
    ticketing strategy and payment channel
        evaluation, 164–65
    transportation governance system,
        168–69
    travel value analysis, 170
    urban automation analysis, 171
    variable tolling, 164
User experience-related questions, 27–29
User-focused mobility services
    analytics, 131
    assumed configuration for, 258
    benefit estimate, 266
    cost estimate, 253
    cost estimation, 245–46
    portfolio of choices, 91
    vision element, 84

Variability, big data, 40
Variable speed limits (VSLs)
    benefits, 199
    defined, 198
    evaluation of, 209
    implementation, 200
    objectives, 199–200
    project, 198–200
    signs, 208
    See also Freeway speed variability
        analysis

Variable tolling use case, 164
Variety, big data, 39–40
Velocity, big data, 36–39
Veracity, big data, 42
Video and light detection and ranging
    (LIDAR) sensors, 67, 69
Video platform, 99
Volume, big data, 35–36

What-how cycle methodology, 7
Whoops moment, 120
Word clouds
    Chapter 1, 2
    Chapter 2, 14
    Chapter 3, 32
    Chapter 4, 56
    Chapter 5, 82
    Chapter 6, 118
    Chapter 7, 138
    Chapter 8, 156
    Chapter 9, 174
    Chapter 10, 196
    Chapter 11, 224
    Chapter 12, 272
    defined, 2
Wow moment, 120